SpringerBriefs in Climate Studies

More information about this series at http://www.springer.com/series/11581

Vladimir Udalov

Behavioural Economics of Climate Change

New Empirical Perspectives

Springer

Vladimir Udalov
School of Business and Economics
University of Wuppertal
Wuppertal, Germany

Dissertation University of Wuppertal, Germany

ISSN 2213-784X ISSN 2213-7858 (electronic)
SpringerBriefs in Climate Studies
ISBN 978-3-030-03531-0 ISBN 978-3-030-03532-7 (eBook)
https://doi.org/10.1007/978-3-030-03532-7

Library of Congress Control Number: 2018962153

Cover Design: eStudio Calamar, Berlin/Figueres

This Springer imprint is published by the registered company Springer Nature Switzerland AG
The registered company address is: Gewerbestrasse 11, 6330 Cham, Switzerland

Acknowledgements

I would like to express my appreciation for those who supported me throughout the last number of years. Firstly, I would like to express my sincere gratitude to my advisor Prof. Dr. Paul J. J. Welfens for his continuous support, patience, motivation and immense knowledge. His guidance helped me at all times while researching and writing this book. Furthermore, he enabled me to participate in numerous international conferences and in the "SINCERE" project funded by the German DFG, where I could exchange my ideas with other researchers.

Besides my advisor, I would like to thank my co-advisor, Prof. Dr. Werner Bönte, for his valuable support, insightful comments and encouragement throughout the research process.

My sincere thanks also go to my co-authors Jens Perret and Véronique Vasseur as well as Prof. Dr. René Kemp for the intensive and productive cooperation. Furthermore, I would like to thank Prof. Dr. Uta Pigorsch, Prof. Dr. André Betzer and indeed all the participants of the Brown Bag Seminar of the Schumpeter School for their many valuable comments and questions.

In addition, I give thanks to everyone who supported me with their advice, encouragement, friendship and patience. Among them my current and former colleagues Niyaz Valitov, Hung Lai, Sonja Jovicic, Christian Dienes, Arthur Korus, Fabian Baier, Nan Yu and all other members of the EIIW. Special thanks also go to David Hanrahan and Valeryia Siarheyeva for their editorial and technical support.

Last but not least, I would like to thank my family for supporting me mentally and emotionally throughout writing of this book and in my life in general.

Wuppertal, Germany
January 2018

Vladimir Udalov

Acknowledgements

Contents

List of Figures

List of Tables

Chapter 1
Introduction

1.1 Motivation

> During the twentieth century we were largely on a comfortable, and a fairly predictable energy path of a mature, fossil-fueled civilization. Things are different now. The world's energy use is at the epochal crossroads. The new century cannot be an energetic replica of the old one and reshaping the old practices and putting in place new energy foundations is bound to redefine our connection to the universe.—Vaclav Smil (2003, p. 373)

The "epochal crossroads", Vaclav Smil refers to, correspond to the profound global environmental changes that have been observed in the past years and decades. Driven by population growth during the second half of the twentieth century and human economic activities, which have been steadily on the rise since the Industrial Revolution, the Earth's climate is changing (Stern 2006).

Indeed, the global economy grew tenfold between 1950 and 2000, while in the same period, the population increases threefold (Jäger 2007). Driven by this development, global energy consumption has increased considerably. According to IEA (2016), CO_2 emissions resulting from the combustion of fossil fuels represent the largest share of global anthropogenic greenhouse gases emissions. Since the Industrial Revolution, annual CO_2 emissions from fuel combustion have increased from near zero to over 32 $GtCO_2$ in 2014 and are still increasing by almost 3% each year. This has led to a significant increase in CO_2 concentrations in the atmosphere over the past century. Whilst during the pre-industrial era the level of CO_2 concentrations accounted for about 280 parts per million (ppm), the average concentration of CO_2 is equal to 399 ppm in 2015, which is about 40% higher than in the mid-1800s (IEA 2016).

Climate experts consider this increase in CO_2 concentrations in the atmosphere to be the main contributing factor in the upward trend of the Earth's surface temperature since 1950 (IPCC 2014). According to Ribes et al. (2017), it is extremely likely that the Earth's climate will indeed become warmer if atmospheric concentrations of CO_2 continue to increase.

V. Udalov, *Behavioural Economics of Climate Change*, SpringerBriefs in Climate Studies, https://doi.org/10.1007/978-3-030-03532-7_1

The consequences of climate change are expected to be substantial and are already being felt. According to IPCC (2014), the atmosphere and ocean have warmed, the amounts of snow and ice have diminished, and the sea level has risen. This development contributes to an increase in the number of extreme weather and climate events. The United Nations Office for Disaster Risk Reduction reports that since 1995 extreme weather events have killed 606,000 people, have affected more than 4 billion individuals in total and have caused 1.8 trillion euros worth of damage to property and infrastructure (UNISDR 2015). It can be concluded that in the case of a further increase in the emission of greenhouse gases, there will be long-lasting changes in all components of the climate system, which will increase the likelihood of severe, pervasive and irreversible impacts for people and ecosystems (IPCC 2014). According to Edenhofer et al. (2013), in order to avoid the risk of dangerous and irreversible climate change, the consensus view is that the global average temperature should not rise above pre-industrial temperatures by more than 2°. Stern (2006, pp. i–iv) states that "the benefit of strong, early action on climate change outweighs the costs and an immediate strong action is required since the costs of stabilising the climate are significant but manageable, whereby delay would be dangerous and much more costly".

Possible strategies to reduce CO_2 emissions can be derived using the Kaya identity, which allows the decomposition of CO_2 emissions into the factors that influence energy consumption and carbon intensity (Kaya 1990):

$$CO_2 = \frac{Y}{P} \times P \times \frac{E}{Y} \times \frac{CO_2}{E}. \tag{1.1}$$

According to Eq. 1.1, CO_2 emissions can be represented as a product of carbon intensity of energy *(CO_2/E)* and factors reflecting the drivers of energy consumption, namely GDP per capita *(Y/P)*, population *(P)* and energy intensity *(E/Y)*. Thus, CO_2 emissions are dependent both on the level of energy consumption and on the makeup of the energy basket. Following Henriques and Borowiecki (2014), CO_2 emissions can be reduced through a lower level of energy consumption, which can be achieved as a result of technological progress, lower economic growth, demographic changes, or by altering the composition of the energy basket to increase the share of sources with lower emission contents.

Driven by the improvements in living standards, the increase in per capita consumption of goods and services as well as the vast increase in population in the past 50–100 years (Swim et al. 2011), CO_2 emissions stem to some extent from the choices people make every day in their roles as consumers and citizens (Berglund and Matti 2006). Having experience with the consequences of climate change, which are widely discussed in the media and academia, means that individuals become more sensitive concerning their own contribution to environmental problems and their possible responses to climate change. Focusing on altering the proximate causes of climate change, individual efforts should consist of such CO_2 mitigation activities that influence energy consumption and carbon intensity (Kaya 1990). These measures might range from a general support of environmental policy to concrete energy saving activities. According to Whitmarsh et al. (2011), on the one hand, individuals can directly reduce their emissions by conserving gas or electricity in the home or by buying electricity from renewable sources. On the other hand, individuals can act indirectly in several roles to promote a low-carbon society. Thus, they can act as low-carbon consumers by buying energy efficient appliances and sustainable products or as low-carbon citizens, for example, by voting for a green policy, joining an environmental campaign or community action group (Whitmarsh et al. 2011).

Individual CO_2 mitigation activities to be investigated in this book are presented in Table 1.1 and can be assigned to the respective components of the Kaya identity.

Taking into account the different components of the Kaya identity, this work considers, on the one hand, individuals' indirect behaviour as a low-carbon citizens, which includes renewable energy support and the choice between economic growth and environmental protection, and, on the other hand, direct individual energy-conserving behaviour in the form of daily energy-saving activities.

Given the wide range of different individual CO_2 mitigation activities to be considered, it must be acknowledged that the individual pro-environmental decision-making process is complex (Berglund and Matti 2006) and can be considered as being guided by a mixture of self-interest and pro-social motivations (Bamberg and Möser 2007). While pro-social motives imply concern for other people, the next generation, other species, or even whole eco-systems (Bamberg and Möser 2007), self-interest motives imply that people tend to protect the environment due to aspects that affect them personally (Stern and Dietz 1994). However, human environmental behaviour does not depend completely on self-interest and pro-social motivations. Various contextual and external factors influence environmental behaviour by either directly affecting behaviour or moderating the relationship between motivational factors and behaviour (Steg and Vlek 2009; Stern et al. 1995).

Taking account of the above-described motives associated with pro-environmental behaviour and decisions, this book investigates the effect of environmental motivations on all types of considered CO_2 mitigation options as well as the impact of non-environmental motives such as economic concern on individuals' renewable energy support. Considering external circumstances, the focus will be on natural disasters and extreme weather events, which, on the one hand, directly influence an individuals' choice between economic growth and environmental protection and, on the other hand, serve as a moderating factor between environmental values and the choice between economic growth and environmental protection as well as an instrumental variable for environmental motivations behind daily energy saving behaviour. The following figure illustrates research questions to be tackled in this book:

Table 1.1 Overview of CO2 mitigation activities

CO_2	=	$\frac{Y}{P} \cdot P$	×	$\frac{E}{Y}$	×	$\frac{CO_2}{E}$
		• Choice between economic growth and environmental protection		• Daily energy-saving behaviour		• Renewable energy support

1.2 Overview

Each chapter of this book deals with one of the selected individual CO_2 mitigation measures, whereby it should be mentioned that each part of this work can be read as an autonomous paper, though they are interconnected as described in Fig. 1.1. In order to give a more detailed overview of the studies discussed in this book, I provide a brief summary of the different chapters.

Chapter 2: Determinants of Individuals' Renewable Energy Support—Empirical Findings for the UK

Chapter 2 investigates the determinants of individuals' renewable energy support. On one hand, individuals' concern regarding short- and long-term increases in electricity prices might have a crucial effect on renewable energy support. Since renewable energy support is often financed by consumers through higher electricity prices and individuals tend to oppose those environmental policies that lead to significant economic costs (Schneider et al. 2010), concern about short-term increases in electricity prices is assumed to have a negative impact on their attitudes towards renewable energy. However, in the long run, electricity market prices might decrease due to potential lower weighted average costs of electricity from renewable energy sources in comparison to estimated fossil fuel-fired electricity generation costs. Thus, individuals' concern about long-term increases in electricity prices is hypothesized to increase their renewable energy support. These non-environmental effects on individuals' renewable energy support are highlighted by the arrow marked with (1a) in Fig. 1.1. On the other hand, since renewable energy support might improve environmental quality, individuals' concern about climate change is considered to have a positive effect of renewable energy support. This effect is captured by the arrow marked with (1b).

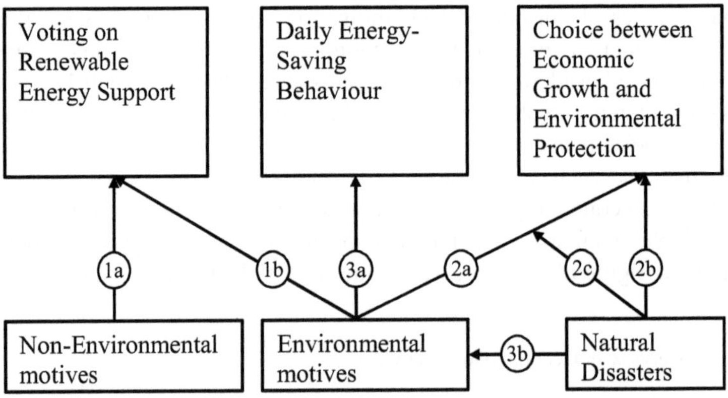

Fig. 1.1 Conceptual framework

As these effects influence population groups to different degrees, especially regarding age structure, renewable energy support might contribute to an intergenerational conflict. That is why, Chap. 2 also investigates the impact of individuals' age on their renewable energy support.

In order to empirically investigate these effects, the 13th wave of the DECC Public Attitudes Tracking survey in the UK is utilised. Empirical results confirm hypothesized effects revealing that respondents who are more concerned about long-term steep rises in energy prices and climate change are more likely to support renewable energy, while individuals who are more worried about paying their energy bills in the short run are less likely to support renewable energy. The empirical analysis also shows that older individuals are less likely to support renewable energy.

Chapter 3: The Impact of Natural Disasters on Individuals' Choice Between Economic Growth and Environmental Protection—Empirical Evidence from the World Values Survey

The purpose of Chap. 3 is to investigate the effect of environmental values and natural disasters on individuals' choice between economic growth and environmental protection, which is illustrated by arrows marked with (2) in Fig. 1.1. While people with inclinations towards environmental values are assumed to be more likely to choose environmental protection (2a), the direct effect of natural disasters on the choice between economic growth and environmental protection (2b) is more ambiguous. On the one hand, individuals might prefer economic growth to environmental protection because natural disasters are associated with capital destruction and imply a decrease in socio-economic security. On the other hand, especially in less affluent countries local environmental problems are compounded when natural disasters strike so that individuals in these countries might opt for environmental protection. Furthermore, a personal experience with a natural disaster can elicit strong emotions, which may contribute to a heightened awareness of climate change risks and reinforce the impact of environmental values on individuals' choice between economic growth and environmental protection (2c).

In order to examine these effects, the 2010–2014 wave of the World Values Survey is combined with data on weather-related natural disasters from the International Disaster Database. The empirical results reveal that for the high-income sub-sample natural disasters have a negative effect on individuals' choice in favour of environmental protection, but increase the impact of environmental values on this choice. As for the low-income sub-sample, natural disasters have no effect on individuals' choice between environmental protection and economic growth nor on the relationship between environmental values and choice in favour of environmental protection. This empirical result might suggest that individuals from less affluent countries do not causally attribute their experience with extreme weather events to climate change (Van der Linden 2015). One possible explanation for this is a lack of knowledge about climate change and its consequences. This result delivers possible policy options for increasing awareness of climate change in developing countries.

Chapter 4: Environmental Motivations Behind Individuals' Daily Energy-Saving Behaviour—Evidence from Germany, The Netherlands and Belgium
Chapter 4 is based on the paper titled "Environmental motivations behind individuals' energy efficiency investments and daily energy-saving behaviour: evidence from Germany, the Netherlands and Belgium" co-authored by Jens Perret and Veronique Vasseur. This paper was published in International Economics and Economic Policy (2017, Vol. 14(3), pp. 481–499) as a part of the special issue on "Resource Efficiency, Circular Economy and Sustainability Dynamics in China and OECD Countries". While Udalov et al. (2017) consider both daily energy-saving activities and energy efficiency investments, Chap. 4 focuses on the impact of environmental motivations (3a) on different types of daily energy-saving behaviour.

The analysis is performed by employing a representative online survey carried out in Belgium, Germany and the Netherlands within the scope of the project "Energy Efficiency of Households in Cities: A Multi-method Analysis" lead by Maastricht University and the Chinese Academy of Sciences. Due to the cross-sectional data design, it is unclear whether environmental motivations cause relevant energy conservation behaviour or energy conservation behaviour causes environmental motivations. This dictates caution in the interpretation of correlations as causal relationships. To deal with this problem, we turn to an instrumental variable analysis. Motivated by using natural disasters as an external factor influencing the relationship between environmental values and the individual's choice between economic growth and environmental protection in Chap. 3, personal experience with extreme weather events serve as an instrumental variable (3b) for environmental motivations behind daily energy-saving behaviour. Considering the estimation results, it can be concluded that apart from the Belgian sub-sample, where environmental motivations are either nonsignificant or even decrease the probability of performing daily energy-saving behaviour, almost all types of daily-energy saving behaviours are positively associated with environmental motivations among the German and Dutch respondents. This result is in line with Stern (1992, 2000) who suggests that pro-environmental actions which are easier to perform are likely to be driven by psychological factors such as environmental motivations.

Chapter 2
Determinants of Individual Renewable Energy Support: Empirical Findings for the UK

2.1 Introduction

In recognition of climate change and global warming, governments across the globe have set targets for reducing carbon emissions, whereby renewable energy provides one of the leading solutions to the climate change issue (IPCC 2011). The problem is that renewable energy technologies are not cost-competitive with conventional technologies, which have benefited for some considerable time from mass production and learning effects (Menanteau et al. 2003). In order to displace the use of fossil fuels, renewable energy technology needs to be promoted with supportive policies, leading to a rapid scale-up of these technologies (Gallagher 2013). As a result, governments utilise a multitude of financial support schemes for renewable energy. However, renewable energy support depends on social acceptance, which is recognized as an important issue shaping the widespread implementation of renewable energy technologies (Moula et al. 2013). Although several empirical studies show high levels of public support for renewable energy technologies (AEE 2016), this might change due to, amongst other things, economic and environmental effects (Akella et al. 2009). Since renewable energy support is financed by the consumers either directly through higher prices for renewable energy or indirectly through taxes (Sundt and Rehdanz 2014; CEER 2017) and individuals tend to oppose such environmental policies that lead to significant economic costs (Schneider et al. 2010), concern about short-term increases in electricity prices might negatively influence individuals' attitudes towards renewable energy. However, in the long run, on the one hand, renewable energy support might improve environmental quality (Panwar et al. 2011) and, on the other hand, decrease electricity market prices due to potential lower weighted average costs of electricity from renewable energy sources in comparison to estimated fossil fuel-fired electricity generation costs (Fürsch et al. 2012; Ciarreta et al. 2012; Akella et al. 2009; IRENA 2015). Anticipating these potential positive effects of renewable energy technologies,

V. Udalov, *Behavioural Economics of Climate Change*, SpringerBriefs in Climate Studies, https://doi.org/10.1007/978-3-030-03532-7_2

individuals' concerns about long-term increases in electricity prices as well as climate change, might contribute to a greater level of renewable energy support.

These effects influence population groups to different degrees, especially regarding age structure. Whereas younger individuals benefit from long-run effects, the group of older individuals faces only a negative short-run effect. Indeed, Jäger and Schmidt (2015) deliver empirical evidence that older individuals tend to discount future payoffs more heavily than working-age individuals showing that there is a negative effect of population aging on public investment such as renewable energy support. Since the main industrial countries are facing the challenge of demographic change, the aging of society might interact with public support for renewable energy, which makes the empirical results also interesting from a policy perspective.

The aim of this chapter is to investigate the impact of individuals' economic and environmental concerns, as well as their age, on renewable energy support. For this purpose the 13th wave of the DECC Public Attitudes Tracking survey (DECC 2015) is utilised. The derived results contribute to Bergmann et al. (2006), who, among other things, estimate the effects of avoiding air pollution, long-term job creation and short- to medium-term increases in electricity prices on an individual's preferences towards renewable energy in Scotland.

The chapter is organized as follows: Sect. 2.2 is devoted to a conceptual framework. Section 2.3 contains the empirical analysis which provides data and variable description and presents the results of ordered logit regressions. Section 2.4 concludes.

2.2 Conceptual Framework

Considering a negative impact of individuals' concern about short-term increases in electricity prices on their attitudes towards renewable energy, it should be noted that wealthier people have more disposable income, have fulfilled their basic material needs and are more interested in investing in post-material goods and services (Inglehart 1995). Indeed, several studies reveal that households with higher income levels have a greater willingness to pay for renewables, for example in the United States, the United Kingdom, Germany, Italy, China and Kenya (Mozumder et al. 2011; Batley et al. 2000; Archtnicht 2012; Abdullah and Jeanty 2011; Liu et al. 2013; Bigerna and Polinori 2014). Since renewable energy support is passed on to consumers, who in the short run pay excessive subsidies in the form of a surcharge on their electricity bill (Sundt and Rehdanz 2014; CEER 2017), renewable energy support reduces disposable income, which, in turn, decreases the willingness to support renewable energy. This result is also supported by Bergmann et al. (2006) who show that short- to medium-term increases in electricity prices reduce respondents' preferences towards renewable energy.

Hypothesis 1 (H1) *Concern about short-term increases in electricity prices has a negative impact on individuals' renewable energy support (Short-term).*

Taking into account a positive effect of long-term benefits from renewable energy support, besides well documented environmental improvements (Panwar et al. 2011), electricity generated by renewable energy sources creates a long-run merit-order effect on power prices and, thus, increases consumers' surplus (Fürsch et al. 2012; Ciarreta et al. 2012, Akella et al. 2009; IRENA 2015).

Furthermore, since individuals with a greater focus on future events are more concerned about the environment (Franzen and Vogl 2013) and environmental concern is correlated with willingness to pay for renewable energy (Ito et al. 2010), individuals who worry about future events such as further climate change or future electricity prices would support renewable energy more strongly. These considerations also go in line with Bergmann et al. (2006), who show that the positive long-term effects of renewable energy such as employment creation and avoiding air pollution are highly valued by individuals.

Hypothesis 2 (H2) *Concern about steep rises in energy prices in the future has a positive impact on individuals' renewable energy support (Long-term).*

Hypothesis 3 (H3) *Concern about climate change has a positive impact on individuals' renewable energy support (Climate Change).*

As regards the negative effect of age on individuals' renewable energy support, Hersch and Viscusi (2006) analyse the impact of age on respondents' willingness to pay higher gasoline prices to address environmental problems. Their empirical results reveal significant age-related differences. For each of the measures of willingness to pay considered, there was a decrease in the willingness to pay value with age. Those over age 65 were half as likely to be willing to pay more for gasoline and, on average, were willing to pay just over one-third as much as were people aged 15–24. Jones and Dunlap (1992, 2002) also show that younger people have a higher willingness to pay for renewable energy. One possible explanation for this is provided by the life-cycle effect (Murphy 1994). Since younger people constitute the closest approximation of future generations, they perceive themselves as being the victim of today's pollution and, thus, are more willing to invest time and energy into solving environmental problems. Furthermore, Jäger and Schmidt (2015) deliver empirical evidence that older individuals tend to discount future payoffs more heavily than working-age individuals showing that there is a negative effect of population aging on public investment such as renewable energy support.

Hypothesis 4 (H4) *Since older individuals unambiguously lose from an increase in renewable energy support, an individual's age has a negative effect on the renewable energy support (Age).*

Figure 2.1 provides an overview over the corresponding hypotheses to be investigated in this chapter.

These hypothesized effects are also investigated and confirmed by Udalov (2014, 2018) who uses a politico-economic overlapping generations (OLG) model provided in Appendix A.

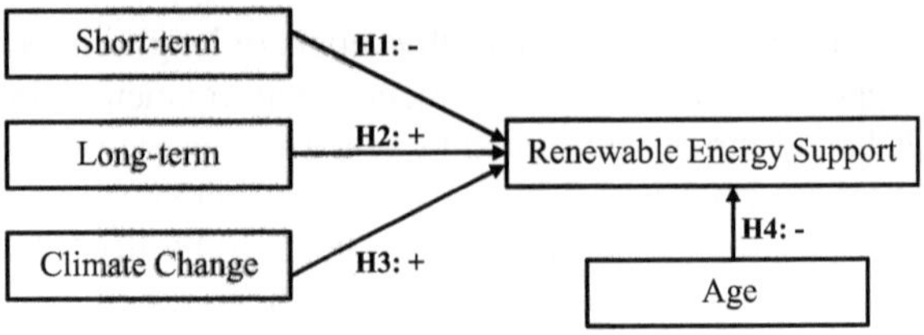

Fig. 2.1 Overview of hypotheses

2.3 Empirical Analysis

2.3.1 Data Description

The 13th wave of the DECC Public Attitudes Tracking survey (DECC 2015) is used to empirically investigate the above formulated hypotheses. The survey was collected between 18 and 29 March, 2015, using face-to-face, in-home interviews with a representative sample of 1981 UK households. The Department of Energy and Climate Change sets up a tracking survey to understand and monitor public attitudes to the energy and environment related issues.

Dependent Variable

The corresponding dependent variable is constructed from individuals' responses to the following question:

- Do you support or oppose the use of renewable energy for providing our electricity, fuel and heat?

Responding to this question, individuals have five alternatives to choose between, ranging from "strongly oppose" (1) up to "strongly support" (5) (Table 2.1).

Independent Variables

Individuals' concern about short- and long-term increases in electricity prices are captured by individuals' responses to the following questions:

- Over the last 3 months, how worried have you been about paying for the energy bills? *(Short-term)*
- How concerned are you about steep rises in energy prices in the future (next 10–20 years)? *(Long-term)*

Thereby, the respondents have four alternatives to choose between, ranging from "not at all worried" (1) up to "very worried" (4) and from "not at all concerned" (1) up to "very concerned" (4).

Individuals' concern about climate change and environmental degradation is captured by considering individuals' responses to the following question:

- How concerned, if at all, are you about climate change, sometimes referred to as "global warming"? *(Climate Change)*

Answer options range also from "not at all concerned" (1) up to "very concerned" (4).

Table 2.1 Dependent variable

Variable	Type	Description	Frequency				
			1	2	3	4	5
RES	Ordinal 1–5	1 if the respondent strongly opposes renewable energy, 5 if the respondent strongly supports renewable energy	32	76	345	889	599

Table 2.2 Independent variables

Variables	Type	Description	Frequency					
			1	2	3	4	5	6
Short-term	Ordinal 1–4	The respondent indicates how worried he or she has been about paying for the energy bills over the last 3 months. 4 if very worried and 1 if not worried at all	530	688	531	205	–	–
Long-term	Ordinal 1–4	The respondent indicates his or her concern about steep rises in energy prices in the future. 5 if very concerned and 1 not concerned at all	102	321	920	621	–	–
Climate change	Ordinal 1–4	The respondent indicates his or her concern about climate change. 4 if very concerned and 1 not concerned at all	181	516	873	380	–	–
Age	Ordinal 1–6	1 indicates that the respondent is between 16 and 24 years old. 6 indicates that the respondent's age is more than 65	264	327	302	323	270	495

In order to investigate the effect of age on individuals' renewable energy support, individuals' responses regarding their age are used, whereby respondents are divided into six age groups, ranging from "16–24" to "older than 65" (Table 2.2).

A vector of control variables includes several socio-economic characteristics that might also be relevant for individuals' renewable energy support. Whereby gender is a widely investigated characteristic. The expectation is that women are willing to pay less due to their lower incomes and, consequently, males and females value differently the costs and benefits related to renewable energy (Bigerna and Polinori 2014). The prosperity hypothesis developed by Diekmann and Franzen (1999) states that an individual's income plays an important role since the quality of the environment is not only a public good but also a good the demand for which rises with income. Following this line of argumentation, income is supposed to have a positive effect on an individual's support of renewable energy. Having children might be also crucial for individual renewable energy support because concern for children's future is a restricted form of social-altruistic attitude (Hansla 2011). Distinctions between rural

Table 2.3 Control variables

			Frequency					
Variables	Type	Description	0	1	2	3	4	5
Female	Binary 0–1	1 if the respondent is female, 0 otherwise	958	1023	–	–	–	–
Income	Ordinal 1–5	The respondent indicates his household's income. 1 if up to £15,999, 5 if more than £50,000	–	615	292	213	182	149
Children	Binary 0–1	1 if the respondent has children, 0 otherwise	1369	612	–	–	–	–
Urban	Binary 0–1	1 if the respondent lives in the urban area, 0 otherwise	313	1658	–	–	–	–

and urban populations are also well documented in the environmental sociology literature since rural and urban locations may exert different influences on participation in environmentally supportive behaviour such as renewable energy support (Huddart-Kennedy et al. 2009). Thus, the following controlling variables are included (Table 2.3).

2.3.2 *Results*

I run several sets of ordered logit specifications on the pooled sample of individual responses. In the first model specification, I include only variables capturing the hypothesized effects associated with individual concerns about increases in electricity prices and climate change. In the second specification, I include additionally all control variables. In order to investigate the H4 hypothesis, I include *Age* into the second model specification as well as age dummies into the third model specification.

Table 2.4 reports the results of the ordered logit regressions. Before proceeding, it should be noted that the estimated coefficient does not reflect the marginal effect on the log odds of renewable energy support. However, its sign provides information about the direction of the effect on the end response categories. Thus, it is possible to interpret the sign and the significance but not the size of the coefficient.

The first estimation result shows that, across all model specifications, individuals who are more worried about paying for their energy bills over the last 3 months are less likely to support renewable energy. The second estimation result refers to the effect of individuals' concern about long-term increases in electricity prices on their renewable energy support. All model specifications indicate that individuals who are more concerned about steep future rises in energy prices are more likely to be in higher categories regarding renewable energy support. The third estimation result reveals that individuals who are concerned about climate change are more likely to support renewable energy.

Table 2.4 Ordered logit estimation results

Explanatory variables	*RES (1)*	*RES (2)*	*RES (3)*
Short-term	−0.215*** (0.053)	−0.228*** (0.062)	−0.233*** (0.063)
Long-term	0.261*** (0.062)	0.269*** (0.074)	0.263*** (0.073)
Climate change	0.631*** (0.056)	0.656*** (0.068)	0.653*** (0.067)
Age		−0.163*** (0.034)	
Age2			−0.179 (0.207)
Age3			−0.266 (0.207)
Age4			−0.356* (0.200)
Age5			−0.481** (0.213)
Age6			−0.886*** (0.195)
Income		0.131*** (0.038)	0.117** (0.039)
Female		−0.355*** (0.104)	−0.349*** (0.104)
Children		−0.089 (0.118)	−0.097 (0.129)
Urban		−0.249* (0.146)	0.253* (0.146)
Pseudo R2	0.045	0.069	0.070
Number observations	1888	1394	1394

Robust standard errors in parentheses
Significance levels: ***$p < 0.01$, **$p < 0.05$, *$p < 0.1$

Since all effects are significant and have the hypothesized signs, the H1, H2 and H3 hypotheses can be accepted. The empirical results go in line with Bergmann et al. (2006).

As regards the effect of age on individuals' support of renewable energy, as hypothesized, age has a significant negative impact on individuals' renewable energy support in the second model specification. In the third model specification, age dummies are additionally included. The second *(25–35)* and the third *(36–44)* age groups show no significant differences in comparison to the first age group *(16–24)* which is considered as a reference group. The fourth age group *(45–54)* shows negative difference to the first age group. This difference stays negative and significant for the fifth *(55–64)* and sixth *(65+)* age groups. Whereby differences to the first age group become more significant the higher the group rank is. This result implies that if individuals stem from the fourth *(45–54)*, fifth *(55–64)* or sixth *(65+)* age groups, they are less likely to support renewable energy compared to the reference group. Thus, the H4 hypothesis that age has a negative impact on the individual's renewable energy support, can also be accepted.

As regards other control variables, income, the key variable from the perspective of the affluence hypothesis, has a consistent positive and significant effect. Male respondents are more likely to support renewable energy. Having children has no significant effect on renewable energy support. Interestingly, people who live in urban areas are less likely to support renewable energy. However, the direction of this effect changes after including age dummies.

Finally, it must be noted that the cross-sectional design of the research dictates extreme caution in the interpretation of correlations as causal relationships.

2.4 Conclusion

Using the 13th wave of the DECC Public Attitudes Tracking survey, conducted by the Department of Energy and Climate Change, this chapter shows that across all model specifications, individuals who are concerned about steep rises in energy prices in the long run are more likely to support renewable energy, while individuals who are worried about paying for the energy bills in the short run are less likely to support renewable energy. Taking into account potential environmental improvements caused by renewable energy, individuals who are concerned about climate change are more likely to support renewable energy. The empirical analysis also reveals that older individuals are less likely to support renewable energy. These results go in line with Bergmann et al. (2006) who, using the applied choice experiment in Scotland, could show that long-term employment creation and avoiding air pollution due to renewable energy are highly valued by respondents, while an increase in electricity prices over the short to medium term reduces consumers' preferences towards renewable energy.

The derived results of this chapter could also be useful from a policy perspective as well. Since older individuals are less likely to support renewable energy, information campaigns might be employed to address the fact that at least the descendants of elderly people would benefit from renewable energy in the long run. Furthermore, since concern about steep rises in energy prices in the long run and climate change positively affects individual renewable energy support, policy makers should increase knowledge and perception of long-term effects of renewable energy support especially among younger individuals, who might directly benefit from these effects, by, for example, using awareness campaigns as well introducing environmental education into the school curricula. However, it should be acknowledged that the cross-sectional design of the research dictates extreme caution in the interpretation of correlations as causal relationships.

Chapter 3
The Impact of Natural Disasters on Individuals' Choice Between Economic Growth and Environmental Protection: Empirical Evidence from the World Values Survey

3.1 Introduction

The warming climate globally contributes to an increase in extreme weather events (Peterson et al. 2012). The United Nations Office for Disaster Risk Reduction reports that since 1995 extreme weather events have killed 606,000 people, have affected more than four billion individuals in total and have caused 1.8 trillion euros worth of damage to property and infrastructure (UNISDR 2015).

Combining the World Values Survey (WVS) with data on natural disasters from the International Disaster Database (EM-DAT), this chapter empirically investigates the impact of climate change-related natural disasters on individuals' choice between economic growth and environmental protection. Insights from behavioural economics and psychology suggest that individual experiences of natural disasters can affect preferences such as those towards environmental protection (Drews et al. 2018) and, thus, may influence public support for environmental policies (Bostrom et al. 1994). This makes the subject of the present analysis worthy of investigation and the results interesting from a policy perspective.

Public opinion regarding the zero-sum choice between economic growth and environmental protection reflects people's attitudes and influences the social-political feasibility of policy implementation (Drews et al. 2018). However, it should be highlighted that the trade-off between economic growth and environmental protection is controversial (Den Butter and Verbruggen 1994) and is not fixed along a country's development path (Grossman and Krueger 1995; Dasgupta et al. 2002). On the one hand, environmental protection and sustainable development might promote economic growth and vice versa. For instance, long-term changes in productivity growth could occur if environmental policies provide permanent incentives to innovate (Kozluk and Zipperer 2014). On the other hand, environmental protection and economic growth are often portrayed as being in conflict with one another. An increase in economic activity is seen as being inevitably bad for the environment, while environmental policy is regarded as imposing a drag on growth.

V. Udalov, *Behavioural Economics of Climate Change*, SpringerBriefs in Climate Studies, https://doi.org/10.1007/978-3-030-03532-7_3

Indeed, at least in the short to medium term, environmental policies such as environmental taxes and tradable emission permits as well as technology-based standards and restrictions impose an additional production-related cost and limit the range of profitable production technologies available (Kozluk and Zipperer 2014).

Since the costs of environmental policies are often borne by consumers, individuals tend to oppose such environmental policies that lead to significant economic costs (Schneider et al. 2010). However, individual experiences with natural disasters can affect an individual's support for environmental policy such as the choice between economic growth and environmental protection. On the one hand, experience with the negative impacts of natural disasters might contribute to a higher awareness of the causes and consequences of climate change, and to the extent to which individuals regard climate change as harmful to their well-being (Brody et al. 2008). On the other hand, natural disasters are also responsible for immediate losses of wealth (Guimaraes et al. 1993), which makes affected individuals opt for economic growth. However, considering less affluent countries in particular, environmental problems in these countries might be compounded when natural disasters strike so that individuals in these countries would choose environmental protection in order to overcome objective local environmental problems (Inglehart 1995).

The chapter is organised as follows: Sect. 3.2 presents the corresponding conceptual framework. Section 3.3 contains the empirical analysis which provides a description of the data and variables and presents the results of cross-sectional regressions based on the total sample as well as high- and low-income sub-samples. Section 3.4 concludes.

3.2 Conceptual Framework

According to Cassar et al. (2017), individual experiences with natural disasters can affect individuals' choice between economic growth and environmental protection thorough various channels.

One possible channel could be through a negative shock to wealth or income. Indeed, since natural disasters are responsible for immediate losses of wealth (Guimaraes et al. 1993), and have a negative effect on economic growth (Raddatz 2007; Hallegatte and Dumas 2009; Noy 2009; McDermott 2012; Klomp and Valckx 2014; Lazzaroni and van Bergeijk 2014), they contribute to a decrease in socio-economic security. Following Inglehart's post-materialism hypothesis (Inglehart 1971, 1997), according to which people tend to embrace less post-materialistic attitudes as socio-economic security decreases, individuals affected by weather-related natural disasters would prefer economic growth to environmental protection. Furthermore, increased perceptions of the probability of being affected by similar disasters in the future may contribute to higher preferences for benefits which accrue earlier rather than later and, thus, lead to a lower propensity to opt for environmental protection, which would prevent weather-related natural disasters in the long term (Cassar et al. 2017).

However, the effect of natural disasters on individuals' choice between economic growth and environmental protection may be conditioned upon the economic setting in which citizens live (Running 2012). Although high-income areas in Europe, North America, and Japan are as highly exposed to natural hazards as are low-income areas in Africa and Asia, high-income societies with efficient and accountable governments can better afford measures to limit the effects of natural disasters (Strömberg 2007). Thus, the impact of natural disasters is particularly detrimental for developing countries. Following the "objective problems, subjective values" (OPSV) hypothesis suggested by Inglehart (1995), which postulates that concern for the environment in less affluent societies might follow from the necessity to overcome objective local environmental problems, one may argue that local environmental problems, particularly in developing countries, are compounded when natural disasters strike so that individuals in these countries would opt for environmental protection. Respondents in more advanced economies are less occupied with severe local environmental problems, since many industrialised countries have prevention measures in place to reduce the risk of severe environmental damage (Ferrier and Spickett 2007). This makes a separate consideration of high- and low-income countries necessary.

Since personal experience of a natural disaster can elicit strong, vivid and memorable emotions (Van der Linden 2015), another possible effect caused by personal experience with natural disasters refers to a higher awareness of climate change, its causes and consequences. Several papers (Brody et al. 2008; Egan and Mullin 2012; Leiserowitz et al. 2012, 2014; Menioux and Zumsteeg 2012; Whitmarsh 2008; Dai et al. 2014) show that those individuals who were personally affected by various kinds of extreme weather events in different regions are more convinced that global warming is a scientific fact, believe in climate change, have a heightened awareness of climate change risks, and are more concerned about environmental problems.

In order to take this effect caused by natural disasters into account, the Attitude-Behaviour-Context (ABC) model suggested by Stern et al. (1995) is used. So far the ABC model has been applied in investigating various pro-environmental behaviours such as recycling (Stern et al. 1995; Hage et al. 2009), participation in green electricity programmes (Clarc et al. 2003), transport choice (Collins and Chambers 2005) and energy saving (Costa and Khan 2013). According to the ABC model, actions or behaviours (B) are associated with internal factors (A) and external conditions (C). Internal factors might be values (Stern and Dietz 1994; Stern et al. 1995; Stern et al. 1999; Stern 2000), norms (Schwartz 1977; Schwartz and Howard 1981) or attitudes (Dunlap and Van Liere 1978; Dunlap et al. 2000). Within this study environmental values are used as internal factors because, in comparison to attitudes and norms, values are relatively few in number, are not situationally specific and are relatively stable (Rokeach 1973). While environmental values serve as positive predictors of individuals' choice to prefer environmental protection (Stern and Dietz 1994), the critical element of the model is that the effect of environmental values on individuals' choice depends on the values of internal and

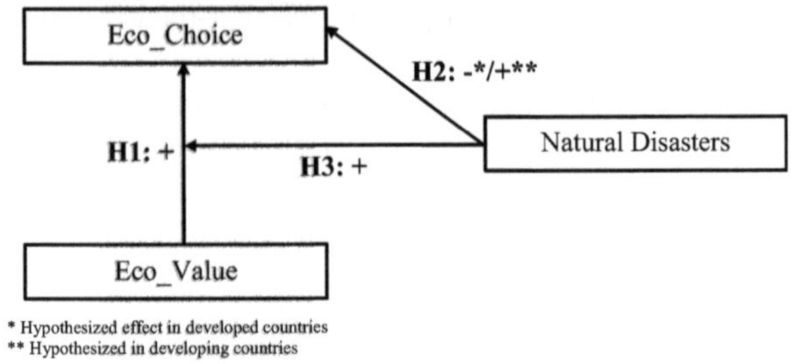

Fig. 3.1 Overview of hypotheses

external factors relative to each other. Considering individuals' choice between economic growth and environmental protection as a behavioural outcome, disasters as an external factor and environmental values as an internal factor, one may argue that natural disasters reinforce the impact of environmental values on individuals' choice between economic growth and environmental protection.

Based on the conceptual framework provided in this section, the following hypotheses are formulated and illustrated in Fig. 3.1.

Hypothesis 1 (H1) *Environmental values (Eco_Value) have a positive effect on individuals' choice to prefer environmental protection to economic growth (Eco_Choice).*

Hypothesis 2 (H2) *While natural disasters have a negative impact on individuals' choice to prefer environmental protection to economic growth in developed countries, they have a positive effect in less affluent countries.*

Hypothesis 3 (H3) *Natural disasters increase the impact of environmental values on individuals' choice to prefer environmental protection to economic growth.*

Figure 3.1 illustrates the corresponding hypotheses to be investigated in this chapter.

3.3 Empirical Analysis

For the empirical investigation, the World Values Survey *(WVS)* is combined with data on natural disasters at a local level from the International Disaster Database *(EM-DAT)*. The 2010–2014 wave of the WVS with 66,278 survey responses across 46 countries, 25 of which are classified by the IMF (2014) as low-income countries, is designed to be a representative survey carried out using consistent methodologies across numerous countries and focusing on changes in the beliefs, values and motivations of people throughout the world. The WVS employs a probabilistic sample method and uses minimum sample sizes of 1000 respondents (Israel and Levinson 2004).

EM-DAT is a global dataset that currently includes around 9700 natural disasters from 1900 to the present day. To be recorded in the database, an event must fulfil at least one of the following conditions: (a) ten or more people reported as killed; (b) 100 people reported as being affected; (c) a state of emergency has been declared; or (d) the country has issued a call for international assistance. EM-DAT includes both natural and man-made disasters (Neumayer and Plümper 2008).

In order to take into account the geographical proximity of individuals to the site of the disaster, catastrophic events at a local level are considered, since they might have a direct effect on the individual's choice between economic growth and environmental protection by eliciting strong emotions, making them more memorable and dominant in processing (Van der Linden 2015). Within the context of the WVS, respondents indicated an exact location (i.e. city/region) where the interview was conducted. This information is used to collect the data from the EM-DAT on the kind, number and magnitude of natural disasters which occurred in the corresponding location. However, it should be acknowledged that for most interviewed individuals, the location where the interview was conducted is assumed to be equal to their place of residence for at least the considered period of 10 years prior to the WVS.

In a first step, I run regressions on the total sample. However, the effect of natural disasters on an individual's choice between economic growth and environmental protection may depend on the country's economic development (Running 2012). On the one hand, individuals living in richer countries tend to embrace more post-materialistic attitudes, while, on the other hand, concern for the environment in developing countries might follow from the necessity to overcome objective local environmental problems (Dorsch 2014). That is why the sample is split into high- and low-income sub-samples using the IMF classification (IMF 2014).[1]

3.3.1 Dependent Variable

My dependent variable is the individual's choice between economic growth and environmental protection. In the survey, respondents were asked to choose between the following statements:

- Protecting the environment should be given priority, even if it causes slower economic growth and some loss of jobs.
- Economic growth and creating jobs should be the top priority, even if the environment suffers to some extent.

[1]Low-income sub-sample: Algeria, Azerbaijan, Armenia, Belarus, China, Colombia, Egypt, Ecuador, Ghana, Iraq, Kazakhstan, Jordan, Lebanon, Libya, Malaysia, Mexico, Morocco, Nigeria, Pakistan, Peru, Philippines, Tunisia, Ukraine, Uzbekistan, Yemen.

High-income sub-sample: Australia, Chile, Cyprus, Estonia, Germany, Japan, South Korea, Kuwait, Netherlands, New Zealand, Poland, Qatar, Romania, Russia, Singapore, Slovenia, Spain, Sweden, Turkey, United States, Uruguay.

Interviewees' responses are used in order to construct the dependent dummy variable *(Eco_Choice)*, which is coded "1" if the individual agrees with the first statement that protecting the environment should be given priority, even if it causes slower economic growth and some loss of jobs. The variable is coded "0" if the respondent agrees with the second statement that economic growth and creating jobs should be the top priority, even if the environment suffers to some extent.

3.3.2 Independent Variables

Natural Disasters

In general, a disaster is defined as being an unforeseen event that causes great damage, destruction and human suffering, overwhelms local capacity and necessitates a request to the national or international level for external assistance (CRED 2015). Since I hypothesize that natural disasters contribute to a heightened awareness of climate change risks and increase the impact of environmental values on an individual's choice in favour of environmental protection, I focus on climatological, hydrological and meteorological disasters, which are consequences of global warming and climate change. In order to take into account the geographical proximity to the disaster site, I consider natural disasters at the local level, where the corresponding WVS interviews were conducted.

Noy (2009) states that whether a disaster event affects the national economy in any given year is likely to depend, on the one hand, on the relative magnitude of the disaster and, on the other hand, on how much time has elapsed since the event took place. Regarding the relative magnitude of a natural disaster, Neumayer and Plümper (2008) argue that the number of people killed *(DKIL)* is the most suitable proxy of disaster strength because it is far less arbitrary than the accounts of the number of people affected or total economic damage, which are much more difficult to estimate and vary across different sources. Following Noy (2009), I standardize my disaster measure for the number of people killed per 1000 inhabitants in the corresponding local area.

Furthermore, natural disasters affect the individual's choice between environmental protection and economic growth beyond the year in which they occur (Noy 2009). On the one hand, the longevity of the effect varies with the severity of the experience with more severe damage or trauma leaving a deeper and longer lasting imprint on affected individuals. On the other hand, natural disasters have short- and long-term effects on economic growth. This might influence the decision to sacrifice economic growth for the sake of environment even years after the natural disaster occurred.

For the sake of simplicity, I consider weather-related natural disasters in the 10-year period prior to the WVS. Since it is likely that a disaster, which occurred in the year when the WVS was conducted has a bigger impact on the individual's choice between environmental protection and economic growth than a disaster that occurred 10 years before the survey was conducted, more recent disasters get a higher weight. Thus, the disaster measure corresponds to

$$Disasters_r = \sum_{n=0}^{9} (10 - n) \frac{DKIL_{r,t-n}}{POP_{r,t-n}} \cdot DNUM_{r,t-n}, \quad (3.1)$$

whereby $DNUM_{r,t-n}$ is the number of natural disasters at the local level in the period *t-n*. $DKIL_{r,t-n}/POP_{r,t-n}$ corresponds to the number of people killed per 1000 inhabitants in the respective local area in the period *t-n*. *(10-n)* is the weighting factor, which decreases, if the time lag *n* between the survey and the corresponding year increases.

Pro-environmental Values

Rokeach (1973) postulates that values are guides for behaviour and have a measurable influence on behavioural choice. The Schwartz theory of basic human values identifies ten basic personal values, which are distinguished between values oriented toward the pursuit of self-interest and values related to a concern for the welfare of others. The universalism value relates to environmental issues since it postulates that individuals may realize that failure to protect the natural environment will lead to the destruction of the resources on which life depends (Schwartz 2012). The World Values Survey captures the environmental dimension of the universalism value by asking respondents to indicate whether they agree with the below description:

- Looking after the environment is important to this person; to care for nature and save life resources.

The variable *Eco_Value is* measured on a six-point scale by taking on the value "6" if the respondent completely agrees with the above description and the value "1" if the respondent completely disagrees with the above description.

Control Variables

I control for several socio-economic and attitudinal characteristics that might be relevant for the choice between economic growth and environmental protection.

Numerous empirical studies provide evidence that women would make more sacrifices for the sake of the environment than men (Zelezny et al. 2000). In particular, women may have a higher willingness to pay for services, which relate to family health. Since environmental degradation increases health risks, women might prefer environmental protection to economic growth (Adebo and Ajewol 2012).

Pro-environmental choice depends on an individual's knowledge of environmental issues. Since knowledge is usually acquired through education, an individual's education level should be positively linked to the *Eco_Choice* (Franzen and Meyer 2010). As regards respondent's age, most studies report its negative effect on *Eco_Choice*. One possible explanation for this is provided by the life-cycle effect theory, which states that younger people perceive themselves as being the victim of today's pollution (Murphy 1994). That is why they are more willing to make sacrifices to solve environmental problems. Following this line of argumentation, having children might also be crucial for the *Eco_Choice* because parents might see their children as being potential victims of today's pollution and thus prefer environmental protection to economic growth (Hansla et al. 2008).

Following the prosperity hypothesis developed by Diekmann and Franzen (1999), the individual's satisfaction with their financial situation plays a crucial role for *Eco_Choice* since the quality of the environment is not only a public good but is also a good the demand for which rises with income. Strongly associated with the prosperity hypothesis is Inglehart's post-materialism hypothesis. According to this, members of wealthy societies are much more willing to give a high priority to protecting the environment. Inglehart's post-materialism index (Inglehart 1995) ranges from 0 to 5, with higher values indicating stronger post-materialistic value orientations. According to Meyer and Liebe (2010), the level of trust might positively affect *Eco_Choice* because trust turns people into unconditional co-operators or makes conditional co-operators confident that others also contribute to public goods.

Because individual-level responses are pooled across countries, unobservable cultural or geographic differences are considered by including country dummies. At the country level, I also control for per capita GDP from 2010 in constant 2005 US dollars, taken from the World Bank.

Since the last wave of the World Values Survey was carried out between 2010 and 2014, the Fukushima nuclear disaster, initiated primarily by the tsunami following the Tōhoku earthquake on 11 March, 2011, might have an effect on individuals' choice between economic growth and environmental protection. For this reason, I control for whether the survey was carried out before or after the Fukushima nuclear disaster. It should be noted that all interviews among respondents from the low-income sub-sample were conducted after the Fukushima nuclear disaster. However, time distance to the Fukushima disaster varies among low-income countries and thus might have an effect on *Eco_Choice* (Table 3.1).

3.3.3 Empirical Strategy

Since my corresponding dependent variable is binary, I apply a common binary probit[2] model on the pooled sample of individual responses. I consider the following baseline specification:

[2]I choose a probit rather than logit model because an interpretation of effects as changes in the probability in a probit regression is more convenient when compared to a logit model where effects are interpreted in terms of odds ratios.

Table 3.1 Variable descriptions and summary statistics

Variables	Type	Description	Mean	SD
Dependent				
Eco_Choice	Binary 0–1	1 if the respondent prefers protecting the environment, even if it causes slower economic growth and some loss of jobs, 0 otherwise	0.53	0.49
Explanatory				
Disasters	Ordinal 0–4.41	Number of natural disasters between 2000 and 2010 weighted by number of people killed and time elapsed data stem from the international disaster database	0.06	0.20
Eco_Value	Ordinal 1–6	The respondent indicates looking after the environment is important to this person, 6 indicates the highest agreement	4.50	1.26
Control				
Age	Ordinal 1–9	1 indicates that the respondent is between 16 and 20 years old. 9 indicates that the respondent's age is between 91 and 100	3.71	3.72
Female	Binary	1 if the respondent is female, 0 otherwise	0.53	0.49
Education	Ordinal 1–9	The respondent indicates the highest educational level that he or she have attained. 9 indicates university-level education with degree	5.77	2.42
Income	Ordinal 1–10	The respondent indicates how satisfied he or she is with the financial situation of his or her household. 10 indicates the highest satisfaction	5.96	2.47
Children	Binary 0–1	1 if the respondent has children, 0 otherwise	0.69	0.45
Trust	Ordinal 1–4	The respondent indicates how much confidence he or she has in environmental organizations. 4 indicates the highest level of confidence	2.61	0.87
Postmat.	Ordinal 1–4	Index based on 12 WVS questions, with higher values indicating stronger postmaterialist value orientation	1.92	1.17
GDP	Cont.	2010 GDP per capita (constant 2005 US$)	24406.8	21557.6
Fukushima	Ordinal 0–4	0 if the survey was carried out before the Fukushima nuclear disaster 1 if the survey was carried out in 2014, 2 in 2013, 3 in 2012 and 4 in 2011 shortly after the catastrophe	2.68	1.21

$$Pr\left(Eco\_Choice_{irj} = 1\right) = \Phi\left(\beta_0 + \beta_1 Eco\_Value_{irj} + \beta_2 Disasters_r + \beta_3 Disasters_r \times Eco\_Value_{irj} + \sum_{k=n+1}^{t} \beta_k c_{k,irj} + \gamma_j + u_{irj}\right), \tag{3.2}$$

where *Eco_Choice* are the binary responses regarding the choice between economic growth and environmental protection. *Eco_Value* denotes the individual's pro-environmental value. *Disasters* is the corresponding natural disaster measure. *Disasters* x *Eco_Value* is the interaction variable between natural disasters and the individual's environmental value, $\sum_{k=n+1}^{t} c_{k,irj}$ are control variables at the individual and country level, γ_j are country dummy variables and u_{irj} are error terms.

β_1, β_2 and β_3 are the corresponding coefficients of interest throughout the chapter. Coefficients β_1 and β_2 measure the effect of *Eco_Value* and *Disasters*, respectively, on the probability that respondents prefer environmental protection to economic growth. The coefficient β_3 measures the effect of natural disasters on the relationship between *Eco_Value* and *Eco_Choice*. However, the interpretation of β_3 is complicated since the magnitude, sign and statistical significance of the interaction effect in nonlinear models might not be correct. Following Norton et al. (2004), I compute the correct marginal effects of the interaction variable.

3.3.4 Results

This section presents the results of probit estimations based on the total sample and on different sub-samples.

Table 3.2 reports the marginal effects of the explanatory variables on the probability that the respondent prefers environmental protection to economic growth. The model specifications (1a), (2a) and (3a) refer to the probit estimations based on the total sample. In the first model specification, I include only variables of interest and country dummies. In the second specification, I additionally include all control variables. However, due to the multicollinearity problem, country dummies are excluded. The third specification differs from the second through the inclusion of country dummies and the exclusion of the GDP per capita and Fukushima variables.

The first estimation result refers to the relationship between *Eco_Value* and *Eco_Choice*. As expected, *Eco_Value* has a significant positive effect on *Eco_Choice* across all model specifications indicating that individuals for whom looking after the environment is important are more likely to prefer environmental protection to economic growth. Thus, the H1 hypothesis can be accepted for the full sample.

The second estimation result refers to the direct effect of weather-related natural disasters on *Eco_Choice*. Although the sign of the corresponding marginal effect

Table 3.2 Probit estimation results based on full sample

Explanatory variables	*Eco_Choice* (1a)	*Eco_Choice* (2a)	*Eco_Choice* (3a)
Eco_Value	0.059*** (0.002)	0.060*** (0.002)	0.060*** (0.002)
Disasters	−0.011 (0.044)	−0.071 (0.047)	−0.008 (0.045)
Eco_Value X Disasters	0.001 (0.009)	0.019** (0.009)	0.002 (0.009)
Female		0.003 (0.004)	0.010** (0.005)
Age		−0.003 (0.002)	0.002 (0.002)
Income		0.008*** (0.001)	0.004*** (0.001)
Education		0.009*** (0.001)	0.011*** (0.001)
Children		0.018*** (0.006)	−0.009 (0.006)
Trust		0.032*** (0.005)	0.030*** (0.006)
Postmaterialism		0.044*** (0.002)	0.042*** (0.002)
GDP		2.21e-08 (1.39e-07)	
Fukushima		0.022*** (0.002)	
Country dummy variables	Yes	No	Yes
Pseudo R2	0.059	0.030	0.069
Number of observations	55,532	50,646	50,646

Marginal effects calculated at the means of the variables reported. Robust standard errors in parentheses
Significance levels: ***$p < 0.01$, **$p < 0.05$, *$p < 0.1$

indicates a negative impact of natural disasters on *Eco_Choice* across all model specifications, the second hypothesis has to be rejected for the full sample, since the effect of *Disasters* turns out to be non-significant.

The result for the interaction variable between natural disasters and *Eco_Value* shows a significant positive effect for most observations in the second model specification only. However, after including country dummies, this interaction effect is non-significant, which means that the significant interaction effect on *Eco_Choice* from the second model specification is reflective of the effect of country dummies on *Eco_Choice* and indicates that natural disasters have no significant impact on the relationship between *Eco_Value* and *Eco_Choice*. Therefore, the H3 hypothesis has to be rejected for the full sample too. Thus, there is only clear evidence for the H1 hypothesis, while the H2 and H3 hypotheses are not valid for the full sample.

Table 3.3 reports the results of probit estimations based on the high-income sub-sample. The H1 hypothesis can be accepted since *Eco_Value* has a significant positive effect on the pro-environmental choice across all model specifications. Due to the negative effect of *Disasters* on *Eco_Choice*, the H2 hypothesis can also be accepted for the high-income sub-sample. This result indicates negative and significant effects across all model-specifications and implies that respondents from the high-income sub-sample who were affected by weather-related natural disasters are less likely to prefer environmental protection to economic growth. The result for the interaction variable between natural disasters and *Eco_Value* shows a significant positive effect for most observations across all specifications, indicating that weather-related natural disasters increase the impact of *Eco_Value* on *Eco_Choice*. This implies that the positive relationship between an individual's environmental value and the decision to prefer environmental protection to economic growth

Table 3.3 Probit estimation results based on high-income sub-sample

Explanatory variables	*Eco_Choice* (1b)	*Eco_Choice* (2b)	*Eco_Choice* (3b)
Eco_Value	0.083*** (0.003)	0.082*** (0.003)	0.085*** (0.003)
Disasters	−0.206** (0.104)	−0.529*** (0.118)	−0.215** (0.109)
Eco_Value X Disasters	0.047** (0.022)	0.066** (0.025)	0.051** (0.023)
Female		0.002 (0.007)	0.003 (0.007)
Age		−0.009** (0.002)	−0.002 (0.002)
Income		0.001 (0.002)	0.0003 (0.002)
Education		0.013*** (0.002)	0.014*** (0.002)
Children		0.010 (0.009)	−0.008 (0.009)
Trust		0.060*** (0.008)	0.061*** (0.009)
Postmaterialism		0.065*** (0.003)	0.065*** (0.003)
GDP		3.26e-08 (1.75e-07)	
Fukushima		0.008** (0.003)	
Country dummy variables	Yes	No	Yes
Pseudo R2	0.059	0.058	0.083
Number of observations	23,430	20,470	20,470

Marginal effects calculated at the means of the variables reported. Robust standard errors in parentheses

Significance levels: ***$p < 0.01$, **$p < 0.05$, *$p < 0.1$

convinced that global warming is a scientific fact, believe in climate change, have a heightened awareness of climate change risks, and are more concerned about environmental problems.

Regarding the low-income subsample, weather-related natural disasters have no effect on individuals' choice between environmental protection and economic growth nor on the relationship between environmental values and pro-environmental choice. This may be due to the fact that individuals from developing countries do not causally attribute their experience of extreme weather events to climate change (Van der Linden 2015). One possible explanation for this is a lack of knowledge about climate change and its consequences.

The finding that weather-related natural disasters increase the impact of environmental values on the choice in favour of environmental protection within the high-income sub-sample but have no effect on the respondents from less affluent countries goes in line with Diekmann and Franzen (1999) and Franzen and Vogl (2013) who state that respondents in more wealthy nations tend to have higher environmental concern. The hypothesis that respondents from the low-income sub-sample would opt for environmental protection in order to overcome objective local environmental problems, which are compounded when natural disasters strike, can be rejected.

The empirical findings may provide possible policy implications. Since environmental values are positively associated with individuals' choice in favour of environmental protection across all model specifications in both low- and high-income sub-samples, a strengthening of environmental values through policy intervention can increase public support for environmental policies, whereby information and awareness-raising campaigns might be employed (Steg and Vlek 2009; Van der Linden et al. 2014).

Concerning the low-income sub-sample, there is a need for basic information provision to overcome a lack of knowledge about climate change and its implications for individuals. This information needs to be communicated through channels perceived to be credible and in a manner that is transparent. This could include adapting marketing techniques to create awareness, acceptance and norms in respect of climate change (Lorenzoni et al. 2007).

Regarding the high-income sub-sample, the implementation of environmental policies in the affected regions might benefit from a greater public acceptance, since natural disasters increase the impact of environmental values on the individual's choice in favour of environmental protection. However, there is a need for information provision regarding the actual impact of natural disasters on economic growth as well as risk prevention measures in respect of severe environmental damage and capital destruction as although high-income societies possess measures to limit the effects of natural disasters (Strömberg 2007), respondents affected by weather-related natural disasters are more likely to opt for economic growth.

The limitations of the present study should be acknowledged as well. It must be noted that the cross-sectional design of the research dictates extreme caution in the interpretation of correlations as causal relationships. Moreover, due to the cross-sectional data used for the purposes of this analysis, it can be difficult to disentangle the effect of natural disasters from other possible geographical effects.

Given the above described limitations of the present study, future research is needed. Since any geographic variation in the outcome variable can only be captured by natural disasters, constructing a multilevel model with natural disasters as a fixed effect and country and sub-country region as random effects would be an important task for future research. In order to investigate the robustness of my empirical results, one should use the next wave of the WVS, which will be made available in 2020. This would benefit the analysis, since more countries and additional relevant questions will be included. Taking the cross-sectional design of the WVS and critical assumptions into account, conducting specific additional surveys in regions which were affected by disasters related to climate change could provide a clearer picture of the effects of natural disasters on individuals' choice between economic growth and environmental protection and other environment-related decisions.

Chapter 4
Environmental Motivations Behind Individuals' Daily Energy Saving Behaviour: Evidence from Germany, The Netherlands and Belgium

4.1 Introduction

Since the residential sector is a substantial consumer of energy in every country, in this chapter we focus on individuals' energy use. The residential sector accounts for approximately 20% of the total delivered energy consumed worldwide and is responsible for 17% of global CO_2 emissions (Nejat et al. 2015). According to Brounen et al. (2012), about one-fifth of total global energy demand originates from the requirements to heat, cool and light residential dwellings. Households in Europe account for 21% of the world's total residential energy consumption (EIA 2016). Despite the fact that energy efficiency in the household sector increased by 19% in EU-27 countries over the period 1990–2008, the final household electricity consumption increased by 13% over the same period (EEA 2010). There are many factors, which explain this upward trend in energy consumption, such as an increase in the number of households, greater comfort demanded, and an increase in electrical appliances in homes (Eurostat 2013).

Households can minimize their energy consumption by increasing the energy efficiency of their stock of appliances or by undertaking daily energy-saving activities. If the aim is to encourage households to reduce energy use, it is important to target determinants of energy use and conservation (Abrahamse and Steg 2009). However, motivations that lead households to adopt energy conservation activities are very complex. On the one hand, economic factors, like saving money on energy bills, seem to be the most important factors influencing energy-saving behaviour. On the other hand, environmental motivations and other related factors might also play an important role (Frederiks et al. 2015).

In particular, the environmental motivations behind energy conservation activities might be interesting from a policy perspective (Urban and Ščasný 2012). Bamberg (2003) considers environmental motivations as situation invariant orientation patterns, which remain stable independently of whether a particular type of energy conservation provides returns. They can reduce the unintended negative

V. Udalov, *Behavioural Economics of Climate Change*, SpringerBriefs in Climate Studies, https://doi.org/10.1007/978-3-030-03532-7_4

consequences of improved energy efficiency, such as the rebound effect. Furthermore, due to their cross-situational influences, environmental motivation might result in a spill-over of environmentally-friendly behaviour to different areas (Whitmarsh 2009). Nevertheless, little research is available on what the exact impact of environmental motivations on different kinds of energy-saving activities undertaken by individuals actually is. This chapter focuses on individuals' daily energy-saving activities by investigating the impact of environmental motivations in Germany, the Netherlands and Belgium.

Since it is unclear whether environmental motivations cause relevant energy conservation behaviour or energy conservation behaviour causes environmental motivations, it is problematic to interpret correlations as causal relationships. To correct for these potential endogeneity problems we make use of instrumental variable analysis[1] for those activities and sub-samples that suffer from endogeneity bias. Personal experience with extreme weather events might serve as a possible instrument for environmental motivations because it is assumed to be correlated with environmental motivations but uncorrelated with energy conservation behaviour.

The analysis is performed by employing data collected in the scope of the project "Energy Efficiency of Households in Cities: A Multi-method Analysis" led by Maastricht University and the Chinese Academy of Sciences. The data has been collected in Belgium, Germany and the Netherlands through an online questionnaire.

Similarities and differences between these countries have to be taken into account. While slight differences in behaviour between the three countries exist, they are neighbouring EU countries whose citizens' behaviour has been shaped for many decades by a comparable legislative framework and similar economic developments. They have also been similarly impacted by the EU's environmental acquis as well as the legislation leading up to the implementation of the first environmental action programme of 1973 and concurring follow-up programmes or the inclusion of environmental issues into the EEC treaty as part of the Single European Act of 1987 (EG 1987). All three countries are impacted comparatively by the seventh EAP (European Commission 2013). Differences between the three countries do still exist especially since the countries are geographically of a different size and layout. Additionally, with the German reunification of the early 1990s, German data might be biased insofar as the eastern parts of Germany have experienced different environmental policies and economic developments and thus formed different behavioural patterns and opinions than the western parts (Weidner 1995).

The next section reviews some of the relevant literature and provides a theoretical background. The following section presents empirical analysis by describing data and relevant variables, providing the employed methodology and empirical results. The chapter concludes with a discussion of implications for the literature and policy makers.

[1]Instrumental variable analysis is a method of estimation that is widely used in many economic applications when correlation between the explanatory variables and the error term is suspected.

4.2 Literature Review and Theoretical Background

Households' energy-conserving behaviour includes a wide range of activities. Jansson et al. (2010) make a distinction between energy efficiency investments and curtailments. Energy efficiency investments involve the acquisition of new technologies, low-energy appliances or energy efficient systems that need monetary investments. These types of behaviour substitute capital for energy and involve one-time purchase decisions, which is associated with an initial financial expense and a potential for future savings (Jansson et al. 2010).

Curtailments refer to non-monetary investments that are behavioural changes such as scheduling efforts, turning off lights, cutting down on heating or on air conditioning and switching off stand-by mode. Curtailment behaviour is made on an everyday basis and involve frequent efforts, and often result in discomfort for the actor performing the behaviour (Jansson et al. 2010). While curtailment behaviour is easily reversible, energy efficient investments consist of the retrofitting of homes to achieve permanent conservation and are irreversible (Dillman et al. 1983). Thus, factors driving the demand for reversible and irreversible decisions are likely to be different.

Existing empirical literature identifies three key determinants that influence the energy consumption of individuals: income, the socio-demographic characteristics of individuals or households and attitudinal variables (Ščasný and Urban 2009). Abrahamse and Steg (2009) divide these key determinants into psychological and socio-demographic factors. It should be noted that determinants of households' energy conservation behaviour affect different activities in different ways depending on their type. Taking socio-demographic factors into account, their effect on different types of energy-conservation activities is ambiguous.

While Lee et al. (2013) show that women are more likely to adopt energy-saving practices and are more willing to pay a higher price for energy-efficient light sources, Poortinga et al. (2003) and Sardianou (2007) report no statistical effect of gender on energy efficient behaviour.

Concerning age effects, Sardianou (2007) shows that conservation investments are less likely to be made by older persons because they expect a shorter stream of benefits from energy improvements than other age cohorts. Another explanation is that younger individuals prefer up-to-date technology that is often more efficient, while older households accept their old appliances and replace them more infrequently (Carlsson-Kanyama et al. 2005). However, Guerin et al. (2000) show that age is positively correlated with the energy-saving habitual behaviour. As regards income, Poortinga et al. (2003) provides an empirical evidence that energy efficiency investments are most acceptable for respondents with a high income, while behavioural measures aimed at reducing direct energy use are the least acceptable for those with high incomes. This might be explained by the fact that technical measures often require an initial investment, which might be less problematic for high-income households (Sardianou 2007). Another possible explanation is connected with the fact that day-to-day actions imply decreased comfort while one-time purchase decisions might even increase consumer's comfort. According to Stern and Gardner

(1981), home ownership also prescribes the type of energy conservation behaviour since energy efficiency investments are available to homeowners, whereas curtailments might be the only option for renters.

4.2.1 Environmental Motivations

Since the purpose of this chapter is to investigate the effect of environmental motivations on energy conservation activities, special attention is paid to psychological factors.

While traditional economic theory postulates that human decision-making and behaviour are based on purely rational choices resting on fundamental assumptions aligned with rational choice theory (Frederiks et al. 2015), a growing body of scientific research demonstrates that human behaviour is rarely driven by the rational choice suggested by traditional economic models. Evidence from psychology and behavioural economics shows that consumer actions often deviate systematically from neoclassical economic assumptions of rationality and might be driven by psychological factors such as environmental motivations. The value-belief-norm theory developed by Stern et al. (1999) explains environmental behaviour by suggesting a causal chain of psychological variables, ranging from basic, general values and beliefs to behaviour-specific beliefs and norms. However, considering differences across the corresponding types of energy conservation behaviour, it should be noted that the effect of psychological factors is ambiguous depending on the type of behaviour.

Stern (1992) shows that actions which are easier to perform or are relatively inexpensive are more likely to be driven by psychological factors. On the contrary, high-involvement activities, which incur considerable monetary costs and also require time and planning activity for their implementation, are more dependent on contextual conditions such as economic concern (Stern et al. 1995). Stern (2000) also shows that the impact of psychological factors such as values, beliefs and norms on individuals' energy conservation behaviour is strongest when contextual factors are neutral.

Indeed psychological factors, which also include environmental motivations, have been identified to be successful in predicting curtailment behaviour. According to Jansson et al. (2010), personal norms, experienced as feelings of a moral obligation to act, affect both curtailment activities and low- to medium-involvement purchase decisions, while environmental beliefs in the form of an ascription of responsibility influence curtailment behaviours. Eriksson et al. (2006) as well as Nordlund and Garvill (2003) show that also for willingness to curtail personal car use, there is a strong influence of personal norms. Social norms, defined as an expectation shared by a group, which specifies behaviour that is considered to be appropriate for a given situation, as well as other people's attitudes and behaviour are important determinants of households' energy-saving behaviour (Gardner and Stern 1996; Sardianou 2007; Ek and Söderholm 2010). Positive attitudes towards energy conservation and the environment, developed as a result of cumulative experience and knowledge, are also associated with higher energy savings (Abrahamse and Steg 2009; Ek and Söderholm 2010; Zografakis et al. 2010).

However, the existing literature does not give a clear answer regarding the impact of psychological factors. Fischer (2008) states that daily energy-saving behaviour is related to habitualised decisions that humans make every day based on previous experience and behaviour. Since curtailment behaviour is associated with changing habits and some discomfort on the individual level, it is not driven by environmental motivations because the environment was not a relevant issue to consider at the time the habit was formed. Gatersleben et al. (2002) and Poortinga et al. (2003) also show that energy use may be particularly predicted by socio-demographic variables, while psychological variables have little impact.

Since the impact of psychological factors on individuals' energy-saving behaviour is ambiguous, the purpose of this chapter is to provide a more in-depth analysis of the impact of environmental motivations on different types of daily energy-saving behaviour in Germany, Belgium and the Netherlands. Following Stern (1992, 2000), we hypothesize that daily energy-saving behaviour which is easier to perform and relatively inexpensive is driven by psychological factors such as environmental motivations.

4.2.2 *Extreme Weather Events*

Due to the cross-sectional data design, it is unclear whether environmental motivations cause relevant energy conservation behaviour or energy conservation behaviour causes environmental motivations. This dictates caution in the interpretation of correlations as causal relationships and might be responsible for the existence of an endogeneity bias.

Motivated by using natural disasters as a moderator variable affecting the strength of the relationship between environmental values and an individual's choice between economic growth and environmental protection in the previous chapter, personal experience with extreme weather events can serve as a possible instrument for environmental motivations. A valid instrument must be related to treatment but neither directly nor indirectly related to outcome, except through the effect of the treatment itself (Rassen et al. 2009). Personal experience with extreme weather events thus might theoretically be a valid instrument[2] since it is correlated with environmental motivations but likely to be uncorrelated with daily energy-saving activities (Fig. 4.1).

[2]It can be argued that the variable measuring extreme weather events is unsuitable to be used as an instrumental variable as it stems from the same survey as all other data and is thus indirectly linked to the energy efficiency measures. While this argument holds, it can be argued that the bias resulting from all variables coming from the same survey is rather small as weather events per se are exogenous and identical for interviewees and non-interviewees, and only their classification as 'extreme' might result in any bias at all. However, as this variable remains the most suitable variable available and does not report any significant correlation even considering sub-samples its bias is considered to be negligible.

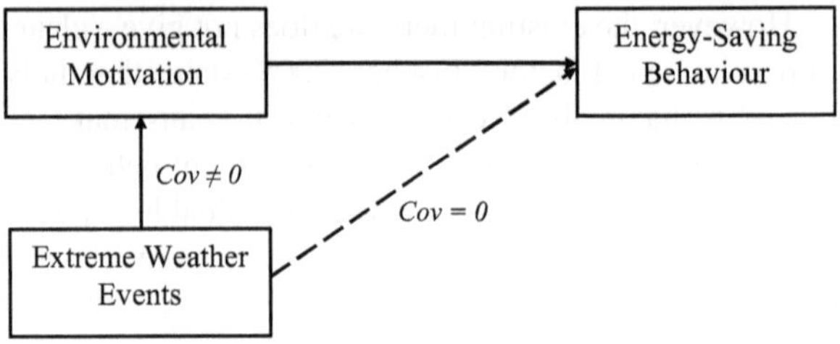

Fig. 4.1 Extreme weather events as an instrument

Considering the literature review from the previous chapter, there is a large body of papers providing evidence that personal experience with extreme weather might be correlated with the environmental motivation. In particular, Brody et al. (2008) states that natural disasters contribute to a heightened awareness of climate change risks and environmental problems. Egan and Mullin (2012) show that weather patterns have a significant effect on people's beliefs about the evidence for global warming. Leiserowitz et al. (2012, 2014) reveal that a large majority of Americans who personally experienced an extreme weather event or natural disaster believe that global warming made a number of recent extreme weather events worse. Whitmarsh (2008) shows that relevant experiences of flooding and air pollution influence individuals' knowledge, attitudes and risk perception regarding climate change. Ménioux and Zumsteeg (2012) show that people who live in areas that are frequently affected by extreme weather events are more convinced that global warming is a scientific fact. Dai et al. (2014) reveal that personal experience with extreme weather events increases global climate change beliefs among Chinese respondents.

Regarding the relationship between extreme weather events and daily energy-saving behaviour, Tsushima et al. (2014) show that after the Great East Japan Earthquake of 2011, most of the interviewed workers felt positive about saving electricity in Japanese offices since a 15% reduction on peak power consumption was required to address the gap between demand and supply capabilities. However, we argue that extreme weather events were not so destructive in Germany, Belgium and the Netherlands in recent years. That is why we assume that experience with extreme weather events are uncorrelated with individual energy-saving behaviours. It should be also noted that the Netherlands has a relatively flat and even countryside and is thus more used to flooding which, particularly in Germany, might already count as a serious natural disaster.

4.3 Empirical Analysis

4.3.1 Data

The analysis employs survey data collected within the scope of the project "Energy Efficiency of Households in Cities: A Multi-method Analysis" led by Maastricht University and the Chinese Academy of Sciences. The survey was carried out in

2016 in the Netherlands, Belgium and Germany and was intended to investigate energy efficiency determinants for passenger transport and energy use in households in cities. The sample size is approximately 400 individuals from each of Belgium and Germany and 450 from the Netherlands. A random sample has been drawn from the population using online survey questionnaires. The corresponding questionnaire can be found in the Appendix B.

4.3.1.1 Description of Variables

The survey collected data on, among others, four different types of daily energy-saving behaviour. In particular, respondents were asked to indicate how often they do the following things: (i) turning the heat down at night, (ii) closing the windows when the heating is running, (iii) turning off the lights and (iv) avoiding leaving appliances on stand-by. Interviewees were asked to give a score from 1 to 5, from never to always. Table 4.1 provides the corresponding descriptive statistics regarding daily energy-saving behaviour. It is remarkable that in all sub-samples the amount of respondents who indicated that they never perform daily energy-saving activities is very low. While the results are generally comparable across the corresponding sub-samples, the most pronounced deviations in this case are Germany with a strong focus on avoiding leaving appliances on stand-by and the

Table 4.1 Descriptive statistics: daily energy saving behaviour

Variables (in %)	1	2	3	4	5
Full sample					
Turning heat down at night	1.29	8.43	10.24	17.30	62.74
Close windows while heating	0.17	4.34	9.96	19.23	66.30
Turning lights off when away	0.34	2.68	7.45	13.57	75.96
No appliances on stand-by	1.71	9.52	15.36	26.33	47.08
Belgium					
Turning heat down at night	1.37	6.56	13.39	19.40	59.29
Close windows while heating	0	2.45	10.87	13.59	73.10
Turning lights off when away	0.26	1.57	9.97	15.22	72.97
No appliances on stand-by	1.96	9.50	17.88	28.21	42.46
The Netherlands					
Turning heat down at night	1.39	3.01	4.86	9.26	81.48
Close windows while heating	0.23	5.77	7.39	20.79	65.82
Turning lights off when away	0.23	3.91	5.75	18.62	71.49
No appliances on stand-by	2.58	15.25	15.25	29.72	37.21
Germany					
Turning heat down at night	1.10	16.76	13.46	24.73	43.96
Close windows while heating	0.27	4.55	12.03	22.99	60.16
Turning lights off when away	0.53	2.38	6.88	6.08	84.13
No appliances on stand-by	0.54	3.53	13.04	20.92	61.96

Netherlands with turning the heat down at night and leaving electrical appliances on stand-by.

The main explanatory variable "*Env_Motiv*" captures individual environmental motivations behind different types of daily energy-saving behaviour and is constructed from responses to the question in which of the following decisions did environmental motivations play an important role: (i) turning the heat down at night, (ii) closing the windows when the heating is running, (iii) turning off the lights and (iv) avoiding leaving appliances on stand-by. Respondents could choose between the following answer options: 0 if "I don't do such things", 1 if "They played no role" and 2 if "They played an important role".

There might be a potential inconsistency caused by respondents who indicate that they do not do such things as daily energy-saving behaviour although they at least occasionally save energy or by interviewees who, on the one hand, state that they save energy due to environmental or non-environmental motivations but, on the other hand, never perform energy-saving behaviour. In order to avoid any biases, we clear the data of any inconsistencies by removing respondents who have shown this answering behaviour. Furthermore, for the sake of a better separation of environmental from non-environmental motivations, we do not consider respondents who indicated that they never perform such actions as daily energy-saving measures. In particular, since the number of individuals that are affected by this is low, and thus the distortion of the original sample remains rather minor, we do not face a significant loss of observations. Taking into account only individuals who curtail energy use because of their environmental and non-environmental motivations allows us to construct a dummy variable "*Env_Motiv*", which is coded "1" if the individual agrees with the statement that environmental motivations played an important role. The variable is coded "0" if the respondent agrees with the statement that environmental motivations played no role. Conditioned that respondents at least occasionally save energy, we are able to measure the effect of environmental motivations on the probability of a more frequent energy-saving behaviour. The corresponding descriptive statistics regarding the environmental motivations behind the respective daily energy-saving activities are provided in Table 4.2.

Interestingly, in the case of Belgium there are significant deviations from Germany and the Netherlands since, in comparison to the other two countries, the share of respondents who indicated that environmental motivations play an important role is significantly lower.

Table 4.2 Descriptive statistics: environmental motivations

Variables:	Full sample		Belgium		Netherlands		Germany	
Environmental motive (in %)	0	1	0	1	0	1	0	1
Turning heat down at night	22.67	77.33	33.80	66.20	18.08	81.92	16.94	83.06
Close windows while heating	23.87	76.13	33.70	66.30	20.14	79.86	18.50	81.50
Turning lights off when away	21.93	78.07	34.74	65.26	18.20	81.80	13.30	86.70
No appliances on stand-by	24.86	75.14	36.75	63.25	20.95	79.05	17.49	82.51

Table 4.3 Descriptive statistics: control variables

	Full sample		Belgium		Netherlands		Germany	
Control variables	Mean	SD	Mean	SD	Mean	SD	Mean	SD
Socio-demographic factors								
Income (1–7)	4.16	1.53	4.23	1.66	4.21	1.41	4.05	1.54
Female (0–1)	0.50	0.50	0.53	0.49	0.47	0.49	0.50	0.50
Age (18–82)	49.00	16.41	47.92	16.43	50.02	16.42	48.95	16.33
Family size (0–8)	1.51	0.94	1.52	1.02	1.38	0.86	1.65	0.93
Education (1–4)	3.38	1.54	2.95	1.02	4.47	1.64	2.59	1.14
Children (0–5)	0.45	0.86	0.41	0.80	0.58	0.95	0.36	0.80
Home related factors								
Home ownership (0–1)	0.48	0.60	0.66	0.47	0.39	0.76	0.42	0.49
Heating space (0–100)	64.78	24.59	64.76	24.26	60.99	24.77	69.05	24.08
Switchable heating (0–1)	0.56	0.49	0.63	0.48	0.59	0.49	0.46	0.50
Comfortable temperature at night (0–40)	16.7	3.60	16.95	3.58	15.65	3.45	17.7	3.47
Comfortable temperature at day (0–40)	17.48	3.02	17.14	2.71	16.76	2.62	18.64	3.37
Psychological factors								
Environmental conscious person (0–1)	0.75	0.43	0.76	0.42	0.72	0.45	0.79	0.41
Worry about environment (0–1)	0.79	0.41	0.81	0.39	0.73	0.43	0.82	0.38
Instrumental variable								
Extreme weather experience (0–1)	0.42	0.49	0.35	0.47	0.39	0.48	0.50	0.50

Considering deviations between the countries reported in Tables 4.1 and 4.2, it becomes imperative to control for the respondents' origins during regression analysis. Since it is difficult to distinguish the effects of environmental motivations from that of other phenomena that potentially influence daily energy-saving behaviour, we include socio-demographic and attitudinal factors as well as dwelling-related factors as control variables.

Table 4.3 indicates that in contrast to the answers above, the control variables are more or less homogenous across the three countries considered, with a slightly larger share of homeowners in Belgium.

4.3.2 Method

In order to explain the impact of environmental motivations, we run ordered probit specifications on the pooled sample of individual responses for each sub-sample and for each type of daily energy-saving behaviour. The baseline specifications that we consider are the following:

$$\Pr(ES\_Behaviour_i > 0) = \Phi\left(\alpha \cdot Env\_Motiv_i + Control_i'\chi + \gamma + u_i\right). \tag{4.1}$$

$\Phi(\cdot)$ is the cumulative distribution function of a normal standard. *ES_Behaviour* is a multiple response dependent variable constructed from responses regarding performing daily energy-saving behaviour. Since *ES_Behaviour* is a polychotomous variable, we apply an ordered probit model. *Env_Motiv* is a dummy variable constructed from responses regarding the role of environmental motivations behind different types of daily energy-saving behaviour and represents specific attitudes that relate directly to the corresponding behaviour. Kaiser et al. (1999) and Stern (1992) state that specific attitudes to energy-related problems and energy saving are better predictors of household energy use than general attitudes. *Control* is a vector of control variables containing socio-economic factors, dwelling-related factors and psychological variables. γ are province dummies and u_i are error terms. The coefficient of interest throughout the paper is α from Eq. (4.1), which measures the effect of environmental motivations on daily energy-saving behaviour.

Due to the fact that it is unclear whether environmental motivations cause relevant energy-saving behaviour or energy-saving behaviour is responsible for environmental motivations, it is reasonable to assume that the direction of causality is ambiguous. To deal with this problem, we turn to an instrumental variable analysis. Personal experience with one or more extreme weather events in the previous few years might provide a possible instrumental variable since it is correlated with environmental motivations but is much less correlated with daily energy-saving activities. After performing a regression-based test for endogeneity, we run an instrumental variable analysis for those activities and sub-samples that suffer from endogeneity. For all other activities and sub-samples, we perform ordered probit regressions.

4.3.3 Results

4.3.3.1 Ordered Probit Estimations

This section presents the results of ordered probit estimations based on the total sample and on different sub-samples. Table 4.4 reports the average marginal effects of the explanatory variables on the probability of being in a higher rather than in a lower category of performing different types of daily energy-saving behaviour for the total sample.

The first estimation result refers to the marginal effects of environmental motivations on various types of energy-saving behaviour. The estimated coefficients of *Env_Motiv* related to "Closing windows while heating", "Turning lights off when away" and "Leaving no appliances on stand-by" are statistically significant in the

Table 4.4 Ordered probit estimation results based on total sample

Explanatory variables	Turning heat down at night	Close windows while heating	Turning lights off when away	No appliances on stand-by
Env_Motiv	−0.008 (0.030)	0.131*** (0.030)	0.090*** (0.028)	0.066** (0.030)
Female	0.065*** (0.025)	0.101*** (0.026)	0.071*** (0.024)	0.086*** (0.0264)
Age	0.003*** (0.001)	0.004*** (0.001)	0.002*** (0.001)	0.003*** (0.001)
Income	−0.005 (0.008)	0.006 (0.010)	0.006 (0.008)	−0.006 (0.010)
Education	0.013 (0.011)	−0.002 (0.010)	0.017* (0.009)	−0.006 (0.011)
Children	0.003 (0.015)	−0.014 (0.016)	−0.013 (0.013)	−0.010 (0.016)
Family size	−0.011 (0.015)	−0.002 (0.015)	−0.014 (0.013)	−0.030** (0.015)
Home ownership	0.026 (0.020)	0.0004 (0.021)	−0.013 (0.017)	0.007 (0.022)
Heating space	−0.001** (0.001)	−0.001** (0.001)		
Switchable heating	0.071*** (0.025)	0.048* (0.026)		
Night temperature	−0.016*** (0.004)	0.006 (0.004)		
Day temperature		−0.026*** (0.006)		
Environmental person	0.047 (0.032)	−0.011 (0.033)	0.063** (0.029)	0.056* (0.034)
Worry about environment	0.104*** (0.033)	0.064* (0.034)	0.057* (0.032)	0.010 (0.036)
The Netherlands	0.291*** (0.036)	0.007 (0.038)	−0.136*** (0.035)	−0.225*** (0.038)
Belgium	0.096*** (0.030)	0.0796** (0.034)	−0.084*** (0.032)	−0.166*** (0.033)
Pseudo R2	0.095	0.054	0.046	0.043
Number of observations	1095	1121	1137	1050

Average marginal effects are reported. Robust standard errors in parentheses
Significance levels: ***$p < 0.01$, **$p < 0.05$, *$p < 0.1$

expected direction. The positive sign of the corresponding coefficients indicates that environmental motivations lead to higher probabilities of performing these types of daily energy-saving behaviour more frequently. The coefficient of *Env_Motiv* associated with “Turning heat down at night” is not significantly different from zero, indicating that environmental motivations do not affect the probability of performing this type of energy-saving behaviour more frequently.

In each ordered probit regression we also include control variables consisting of socio-demographic factors, dwelling-related factors and psychological factors as well as country dummies. Considering socio-demographic factors, respondent's age and sex are statistically significant for each kind of daily energy-saving behaviour and have a positive sign, indicating that older and female respondents are more likely to curtail energy use on daily basis. Besides "Leaving no appliances on stand-by", variables characterising the structure of households such as having children and family size seem to have no significant effect on daily energy-saving behaviour. Interestingly, the coefficient for income is not significantly different form zero, indicating that respondent's income does not affect the probability of daily energy-saving behaviour. Being positively associated only with "Turning lights off when away", respondent's level of education has also a rather limited effect on all other types of daily energy-saving behaviour.

As to the effect of home-related factors, as expected, heating specific variables seem to have a significant impact on "Turning heat down at night" and "Closing windows while heating". Although the existing literature indicates that home ownership prescribes the type of energy-conserving behaviour, our empirical results show that home ownership has no significant effect on daily energy-saving behaviour.

Concerning the effects of attitudinal variables, estimated coefficients indicate that being an environmentally conscious person lead to higher probabilities of performing "Turning lights off when away" and "Leaving no appliances on stand-by", while worries about the environment are positively associated with "Turning heat down at night", "Closing windows while heating" and "Turning lights off when away".

Since there are deviations regarding daily energy-saving behaviour as well as environmental motivations across the considered countries, control for the respondents' origins is imperative. Taking Germany as a baseline sub-sample against which the results for Belgium and the Netherlands are measured, one can identify that respondents' origins have ambiguous effects on the probability of performing different types of daily energy-saving behaviour. While being from Germany increases the probability of "Turning lights off when away" and "Leaving no appliances on stand-by", it decreases the probability of "Turning heat down at night" and has an unclear effect on "Closing windows while heating". The coefficient for the Netherlands is not significantly different from zero, indicating that there are no significant differences between Germany and the Netherlands. However, respondents from Belgium are more likely to close windows while heating in comparison to German respondents.

Taking into account the ambiguous effect of respondents' origins, Table 4.5 presents the coefficient estimates from within country ordered probit regressions for each type of daily energy-saving behaviour. Considering the marginal effects of environmental motivations, empirical results demonstrate that environmental motivations lead to higher probabilities of a more frequent performance of almost all types of daily energy-saving behaviour among German and Dutch respondents. Only in the Belgian sub-sample is the effect of environmental motivations either nonsignificant for most of the energy-saving activities considered or even reduce the probability that individuals turn heat down at night more frequently.

Table 4.5 Ordered probit estimation results based on German, Dutch and Belgium sub-samples

Explanatory variables	Turning heat down at night	Close windows while heating	Turning lights off when away	No appliances on stand-by	Turning heat down at night	Close windows while heating	Turning lights off when away	No appliances on stand-by	Turning heat down at night	Close windows while heating	Turning lights off when away	No appliances on stand-by
	Germany				The Netherlands				Belgium			
Env_Motiv	0.156*** (0.060)	0.235*** (0.053)	0.110*** (0.037)	0.161*** (0.053)	0.062 (0.044)	0.257*** (0.047)	0.207*** (0.053)	0.122** (0.057)	−0.100* (0.053)	0.001 (0.052)	−0.014 (0.052)	0.025 (0.051)
Female	0.100** (0.045)	0.146*** (0.044)	0.108*** (0.034)	0.075 (0.047)	0.037 (0.037)	0.018 (0.042)	−0.024 (0.044)	0.058 (0.046)	0.026 (0.044)	0.084* (0.043)	0.077* (0.043)	0.085* (0.044)
Age	0.001 (0.002)	0.003** (0.001)	0.001 (0.001)	0.002 (0.002)	0.002 (0.001)	0.003** (0.001)	0.001 (0.002)	0.001 (0.002)	0.005*** (0.001)	0.006*** (0.001)	0.005*** (0.001)	0.006*** (0.001)
Income	−0.002 (0.016)	−0.015 (0.017)	−0.005 (0.011)	−0.011 (0.016)	−0.001 (0.014)	−0.004 (0.018)	−0.019 (0.016)	0.001 (0.020)	−0.012 (0.015)	0.009 (0.015)	0.022 (0.015)	−0.007 (0.015)
Education	−0.001 (0.021)	−0.049*** (0.019)	0.009 (0.014)	−0.019 (0.020)	0.022* (0.013)	0.013 (0.015)	0.045*** (0.016)	−0.006 (0.017)	0.016 (0.023)	0.026 (0.022)	−0.014 (0.023)	0.015 (0.024)
Children	0.014 (0.032)	−0.019 (0.028)	−0.048*** (0.015)	−0.031 (0.025)	0.001 (0.018)	−0.005 (0.021)	0.004 (0.022)	0.009 (0.026)	−0.018 (0.027)	−0.030 (0.033)	−0.003 (0.027)	−0.033 (0.032)
Family size	−0.019 (0.030)	−0.013 (0.026)	−0.024 (0.016)	−0.070*** (0.024)	0.008 (0.022)	−0.017 (0.024)	0.005 (0.026)	−0.004 (0.027)	−0.022 (0.024)	0.038* (0.023)	−0.008 (0.024)	−0.022 (0.028)
Home ownership	0.045 (0.053)	0.002 (0.049)	0.009 (0.034)	0.073 (0.051)	0.031 (0.019)	−0.035 (0.023)	−0.035 (0.022)	−0.012 (0.031)	0.074 (0.050)	0.036 (0.047)	−0.021 (0.049)	0.053 (0.053)
Heating space	−0.001 (0.001)	−0.001 (0.001)			−0.001* (0.001)	−0.002** (0.001)			−0.001 (0.001)	−0.001 (0.001)		
Switchable heating	0.012 (0.051)	0.081* (0.047)			0.092*** (0.035)	−0.037 (0.040)			0.053 (0.045)	0.077* (0.043)		
Night temperature	−0.013* (0.007)	0.002 (0.007)			−0.020*** (0.007)	0.012** (0.006)			−0.017*** (0.006)	−0.002 (0.007)		
Day temperature		−0.024*** (0.008)				−0.033*** (0.012)				−0.017* (0.009)		
Environmental person	0.089 (0.059)	−0.034 (0.060)	0.008 (0.039)	0.021 (0.062)	0.024 (0.0407)	−0.018 (0.050)	0.054 (0.048)	0.069 (0.054)	−0.020 (0.066)	−0.045 (0.062)	0.065 (0.062)	0.041 (0.064)

(continued)

Table 4.5 (continued)

Explanatory variables	Turning heat down at night	Close windows while heating	Turning lights off when away	No appliances on stand-by	Turning heat down at night	Close windows while heating	Turning lights off when away	No appliances on stand-by	Turning heat down at night	Close windows while heating	Turning lights off when away	No appliances on stand-by
	Germany				The Netherlands				Belgium			
Environmental worry	0.093 (0.067)	0.050 (0.062)	0.126*** (0.040)	0.081 (0.0654)	0.049 (0.037)	0.022 (0.049)	−0.028 (0.055)	−0.070 (0.052)	0.160** (0.068)	0.103 (0.063)	0.111 (0.068)	0.058 (0.067)
Province dummy	Yes	Yes	Yes	Yes	Yes	Yes	Yes	Yes	Yes	Yes	Yes	Yes
Pseudo R2	0.051	0.108	0.149	0.064	0.101	0.108	0.051	0.025	0.093	0.082	0.074	0.0625
Number of observations	357	370	373	363	397	402	403	352	341	349	361	335

Average marginal effects are reported. Robust standard errors in parentheses
Significance levels: ***$p < 0.01$, **$p < 0.05$, *$p < 0.1$

Tables B1, B2 and B3 in Appendix C provide the necessary robustness checks by running three model specifications for each type of daily energy-saving behaviour and for each country, whereby the first model specification corresponds to that used in the Table 4.5. In the second model specification, we include only "*Env_Motiv*" and province dummies. In the third specification, we include "*Env_Motiv*" as well as control variables but exclude province dummies. Across all model specifications the sign and significance of the main explanatory variable remain the same. The only exception is "Turning heat down at night" in the Dutch sub-sample, which is significantly enhanced by environmental motivations, if we exclude either province dummies or control variables.

As for socio-demographic factors, within-country regressions indicate that they have only a limited effect on daily energy saving behaviour. While the estimated coefficients of the most of socio-demographic variables are not significantly different from zero, the respondent's age is positively associated with all types of energy-saving behaviour in Belgium and with "Closing windows while heating" in Germany and the Netherlands. Female respondents are more likely to perform almost all types of energy-saving behaviour in Germany and Belgium. Although it is expected that behavioural measures aimed at reducing energy use are the least acceptable for high incomes, respondent's income has no significant effect on daily energy-saving behaviour in all sub-samples. Interestingly, respondent's level of education has an ambiguous effect. While it is positively associated with "Turning heat down at night" as well as "Turning lights off when away" among Dutch respondents and even reduces the probability of "Closing windows while heating" in Germany, respondent's level of education has no significant effect on performing energy-saving activities among Belgian respondents. As regards variables characterizing the structure of households, German respondents who have a large family and children are less likely to turn lights when leaving and to leave no appliances on stand-by, respectively.

As in the full sample, the effect of home ownership is nonsignificant across all sub-samples and types of daily energy-saving behaviour. Heating related factors strongly influence "Turning heat down at night" and "Closing windows while heating" in the Dutch sub-sample and have only a rather limited impact on these types of behaviour in Germany and Belgium. Information regarding comfortable temperatures at night and during the day has a significant effect on heating related energy-saving behaviour across all sub-samples.

Regarding attitudinal variables, only in the German and Belgian sub-samples do worries about the environment have a positive effect on "Turning lights off when away" and "Turning heat down at night", respectively. In the remaining sub-samples attitudinal variables do not significantly differ from zero, indicating that they do not affect the probability of a more frequent performing of the relevant daily energy-saving behaviour.

4.3.3.2 Instrumental Variable Analysis

Considering the marginal effects associated with environmental motivations, it should be noted that the results reported in Tables 4.4 and 4.5 have to be interpreted with caution. According to Sabatini (2012), there are two main reasons to suspect the existence of endogeneity problems. First, daily energy-saving behaviour and environmental motivations are individual choices, which depend on individual specific and unobservable preferences. Unobservable individual effects, such as time preferences, personal interests, and individuals' exogenous shocks, may be correlated both with daily energy-saving behaviour and environmental motivations. Second, it is unclear whether environmental motivations cause relevant energy conservation behaviour or energy conservation behaviour causes environmental motivations, which might result in reverse causality and an endogeneity problem. That is why, at first, we have to perform a regression-based test to check whether environmental motivations are endogenous. If the test fails to reject the absence of endogeneity, we can use the results from order probit regressions reported in the Tables 4.4 and 4.5. Otherwise, we are prompted to address endogeneity through IV estimates, whereby the instrumental variable is constructed from responses to the following question: Have you personally experienced one (or more) extreme weather event in the last few years? Respondents could choose between the following answer options: 1 if "Yes" and 0 if "No".

Following Sabatini (2012), a two-stage procedure is used. In the first stage we use the variable "*Env_Motiv*" as the dependent variable and all of the exogenous variables as regressors, i.e., the instrumental variable and all exogenous variables are included in the model. At the second stage, we regress individuals' responses regarding daily energy-saving activities on the predicted residuals from the first stage, on the potential endogenous variable, and on all the exogenous variables. A standard t-test for the predicted residuals is our test for endogeneity. The null-hypothesis is that the effect of predicted residuals from the first stage is zero and that therefore the main explanatory variable is exogenous. Rejecting the null hypothesis thus indicates the presence of an endogeneity problem. In this case an instrumental variable analysis has to be used.

Table 4.6 reports the corresponding effects of the predicted residuals from the second stage on daily energy-saving activities. We reject the null hypothesis that environmental motivations are exogenous for the bold highlighted daily energy

Table 4.6 Test for endogeneity

	Predicted residuals			
Dependent variables	Full sample	Germany	The Netherlands	Belgium
Turning heat down at night	**−0.453*** (0.273)**	**−1.198**** (0.512)**	0.531 (0.534)	0.200 (0.387)
Close windows while heating	**−0.792**** (0.258)**	−0.187 (0.501)	0.166 (0.411)	−0.109 (0.396)
Turning lights when away	−0.350 (0.264)	0.463 (0.631)	−0.219 (0.435)	−0.126 (0.378)
No appliances on stand-by	**−0.479**** (0.242)**	−0.378 (0.594)	−0.177 (0.401)	0.240 (0.351)

Presence of endogeneity is highlighted in bold
Significance levels: ***p < 0.01, **p < 0.05, *p < 0.1

saving activities in the selected samples. It is remarkable that besides "Turning lights off when away" environmental motivations are endogenous for all daily energy-saving activities in the full sample. While in the German sub-sample only environmental motivations associated with "Turning heat down at night" are endogenous, environmental motivations behind all types of energy-saving behaviour are exogenous in the Dutch and Belgian sub-samples.

Taking into account possible endogeneity problems, Table 4.7 reports the results of ordered probit estimations for exogenous environmental motivations and IV ordered probit analysis for endogenous environmental motivations within the full sample as well as for each country. In order to generate the two-stage ordered probit results, the models are estimated using the user-written Conditional Mixed Process (CMP) command in Stata, developed by Roodman (2011). The CMP procedure calculates its estimators from a maximum likelihood approach over a multivariate normal distribution. With this procedure, we are able to take into account the potential endogeneity of a right-hand variable that is not continuous but dichotomous or polychotomous.

For the sake of brevity, we report only the marginal effects of environmental motivations on different types of individuals' daily energy-saving behaviour. However, it should be mentioned that the corresponding control variables are included in every regression as well.

In cases where environmental motivations are endogenous, IV estimates are conducted and highlighted in bold. It should be noted that the strength of our instrument is tested by performing the χ^2-test on the instrument in the first stage regression. The χ^2-test reports significance for all estimated relations for a margin of error at the 10% level. Considering the bold highlighted coefficients reported in Table 4.7, we can identify the following deviations from empirical results provided

Table 4.7 Ordered probit and IV estimation results

	Turning heat down at night	Close windows while heating	Turning lights off when away	No appliances on stand-by
Full sample				
Env_Motiv	**−0.321**** **(0.106)**	**−0.029** **(0.148)**	0.090*** (0.028)	**−0.390***** **(0.056)**
Germany				
Env_Motiv	**−0.419***** **(0.060)**	0.235*** (0.053)	0.110*** (0.037)	0.161*** (0.053)
The Netherlands				
Env_Motiv	0.062 (0.044)	0.257*** (0.047)	0.207*** (0.053)	0.122** (0.057)
Belgium				
Env_Motiv	−0.100* (0.053)	0.001 (0.052)	−0.014 (0.052)	0.025 (0.051)

IV estimates are highlighted in bold
Average marginal effects are reported. Robust standard errors in parentheses
Significance levels: ***$p < 0.01$, **$p < 0.05$, *$p < 0.1$

in Tables 4.4 and 4.5. Concerning the full sample, the marginal effects of environmental motivations on "Turning heat down at night" and "Leaving no appliances on stand-by" become negative and are both significantly different from zero, indicating that environmental motivations even decrease the probability of being in a higher category of performing these types of energy-saving behaviour. In the case of "Closing windows while heating", environmental motivations turn out to be non-significant after running the IV analysis. Within the German sub-sample, environmental motivations behind "Turning heat down at night" was identified to be endogenous. After applying the IV analysis, the marginal effect of environmental motivations remains significant but negatively affects the likelihood of turning heat down at night more frequently. Interestingly, the marginal effects of endogenous environmental motivations estimated using the IV ordered probit analysis are negative and thus imply lower probabilities of performing these types of daily energy-saving behaviour.

As regards the marginal effects of exogenous environmental motivations, they stem from the Tables 4.4 and 4.5 and were already explained in detail.

4.4 Conclusion

This chapter investigated the impact of environmental motivations on individuals' daily energy-saving behaviour in Germany, the Netherlands and Belgium. The corresponding data was collected within the scope of the project "Energy Efficiency of Households in Cities: A Multi-method Analysis" led by Maastricht University and the Chinese Academy of Sciences.

Considering the ordered probit results, it can be concluded that, apart from the Belgian sub-sample, where the effect of environmental motivations is either nonsignificant or even reduce the probability of performing daily energy-saving behaviour, environmental motivations increase the probability of a more frequent performing of almost all types of daily energy-saving behaviour in the remaining sub-samples.

However, these results have to be interpreted with caution since we suspect the existence of reverse causality and endogeneity problems. An instrumental variable approach is used to get around this problem, whereby personal experience with one or more extreme weather events in the last few years serves as an instrumental variable. Applying the IV ordered probit changes the results insofar as marginal effects of environmental motivations turn out to be negative and imply lower probabilities of performing energy-saving behaviour, especially in the full sample. Considering the respective types of energy-saving behaviour, environmental motivations even reduce the probability of "Turning heat down at night" in the full sample as well as among the respondents from the German and Belgian sub-samples.

The empirical findings, which show that environmentally motivated German and Dutch respondents are more likely to curtail energy use on a daily basis, go in line with Stern (1992, 2000) and Stern et al. (1995). They state that daily energy-saving behaviour, which neither incurs considerable monetary costs nor requires time and planning activity for their implementation, is more likely to be driven by psychological factors such as environmental motivations than by non-environmental motivations such as, for instance, economic concerns. Following Urban and Ščasný (2012), this empirical finding might have practical policy implications, since a strengthening of environmental motivations through policy intervention can reinforce daily energy-saving behaviour, whereby awareness-raising campaigns might be employed to stimulate environmental motivations. However, several studies concluded that even if a campaign is very intensive, and uses several forms of media, the effect in terms of new habits and a changed behaviour takes time to register (Dexter 1964; Windahl and Signitzer 1992; Henryson et al. 2000).

Considering the negative effect of environmental motivations on "Turning heat down at night" across almost all sub-samples, this type of energy-saving behaviour seems to be associated with non-environmental motivations. In this case monetary rewards and financial incentives might serve as an extrinsic driving force to save energy (Abrahamse et al. 2005).

Taking into account the Belgian sub-sample, the coefficients of environmental motivations are not significantly different from zero for almost all types of daily energy-saving behaviour. In this case policy makers should anticipate that daily energy-saving behaviour is related to habitualised decisions and ensure that the environment is a relevant issue to consider at the time the corresponding habit is in the process of forming (Fischer 2008). Thus, policy-makers should introduce environmental education into the school curricula. Education programmes should reflect the importance of an ethic for living sustainably. Further research is necessary for analysing the effect of these education programmes. Additional intervention strategies for enhancing daily energy-saving behaviour will be discussed in the next chapter.

These results offer an incentive for a more in-depth analysis of the dynamics of the underlying samples as the implemented data clearly reflects characteristics of data collected by other studies. Future research will also consider Chinese survey results, which will also be collected within the scope of the project "Energy Efficiency of Households in Cities: A Multi-method Analysis".

The limitations of the present study should be acknowledged as well. It must be noted that the cross-sectional design of the research dictates extreme caution in the interpretation of correlations as causal relationships. In order to obtain the causal mechanism, we make use of instrumental variable analysis. However, if there were a direct effect of personal experience with one or more extreme weather events in the last few years on daily energy-saving behaviour, applying an IV approach would lead to bias (Angrist and Krueger 2001). We argue that extreme weather events were not so destructive in Germany, Belgium and the Netherlands in recent years. That is why experience with extreme weather events is weakly correlated with individual energy-saving behaviour. Further, it needs to be recalled that using a variable that has been part of the same survey as all other variables as an instrument might also bias the IV estimation. Thus, using a truly exogenous variable might increase the validity of the IV estimation and strengthen the results of this study.

Chapter 5
Concluding Remarks

Many countries will need to introduce climate mitigation policies over a long period of time in order to help stabilize the climate and avoid further global warming respectively. Climate change stands for a global public good and creates its own international public good challenges, since each country's emission of greenhouse gases contributes cumulatively to the increase of the overall concentration, and each country's abatements entail higher costs than benefits, unless effective concerted and collective efforts take place (Grasso 2004). In the context of the UN Paris climate convention this problem has become more apparent as the US, under the Trump Administration, has announced its intention to withdraw from this global cooperation approach. Claiming that the human contribution to climate change is uncertain, and that the ability to predict the effects is limited, the Trump Administration stands for a sudden shift in US climate policy (Trump 2017). Nevertheless, there is a wide public perception that human activities do indeed cause changes in the Earth's atmosphere and contribute to climate change (IPCC 2013). The influential Stern review on the economics of climate change emphasizes that "the benefit of strong, early action on climate change outweighs the costs and an immediate strong action is required since the costs of stabilising the climate are significant but manageable, whereby delay would be dangerous and much more costly" (Stern 2006, pp. i–iv). In particular, changes in the behaviour and preferences of households and consumers can play a crucial role, since they result—in combination with technological progress and macroeconomic adjustment dynamics—in large reductions in the emission of greenhouse gases. Individuals who have had personal experience with the consequences of climate change, which are widely discussed in both the media and academia, have become more sensitive concerning their own contribution to environmental problems and their possible options regarding the mitigation of climate change, which might range from general support of environmental policy to concrete direct energy-saving activities. Since changes in consumption patterns and preferences can achieve considerable reductions in emissions at relatively low costs (Faber et al. 2012), determinants of the pro-environmental decision-making process require a deeper investigation.

V. Udalov, *Behavioural Economics of Climate Change*, SpringerBriefs in Climate Studies, https://doi.org/10.1007/978-3-030-03532-7_5

Taking into account the variety of possible behavioural climate change mitigation options, this book considered, on the one hand, individuals' renewable energy support as well as the choice of individuals between economic growth and environmental protection, and, on the other hand, direct daily energy-saving behaviour, whereby individual pro-environmental decision-making is supposed to be influenced by a mixture of environmental and non-environmental motivations as well as external circumstances such as natural disasters. Figure 5.1 summarizes the corresponding findings of this work.

Considering environmental motives, they are positively associated with individuals' renewable energy support and choice in favour of environmental protection across all sub-samples. In the case of daily energy-saving behaviour, the marginal effects of environmental motivations are significantly different from zero and have a positive sign in the German and Dutch sub-samples, indicating that environmental motivations lead to a higher probability of energy-saving behaviour. However, among the Belgian respondents, environmental motivations are either nonsignificant or even decrease the probability of engaging in daily energy-saving behaviour. As regards non-environmental motivations, individuals who are concerned about steep rises in energy prices in the long run are more likely to support renewable energy while individuals who are worried about paying for the energy bills in the short run are less likely to support renewable energy. Concerning natural disasters, personal experience with extreme weather events serve as an instrumental variable in order to identify causal effects of environmental motivations on different kinds of individual daily energy-saving behaviours. Furthermore, natural disasters turn out to have an ambiguous effect on individuals' choice between economic growth and environmental protection among respondents from different sub-samples. Thus, within the high-income sub-sample, natural disasters have a negative effect on individuals'

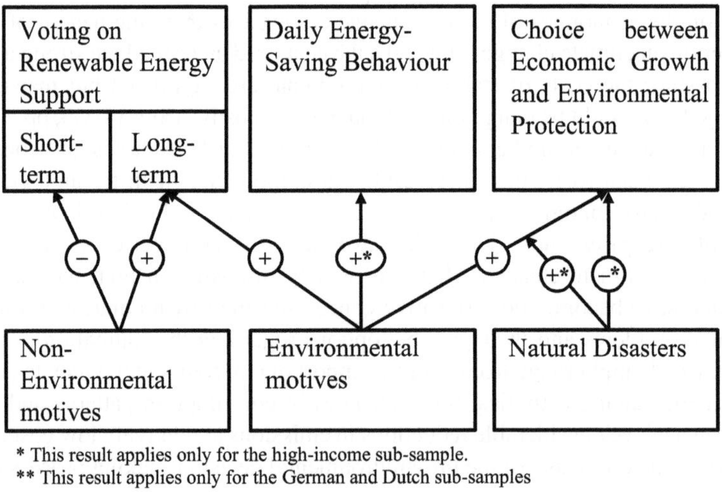

Fig. 5.1 Overview of key findings

choice in favour of environmental protection, but increase the impact of environmental values on this choice. However, within the low-income sub-sample, natural disasters have no effect on individuals' choice between environmental protection and economic growth nor on the relationship between environmental values and pro-environmental choice.

5.1 Policy Implications

The derived results of this book could also be useful from a policy perspective. On the one hand, policymakers in ageing societies in Europe or Asia such as, for instance, Germany, France, Italy, Japan, Republic of Korea and in the long run possibly China, should be interested in the implications derived from the empirical analysis of individual renewable energy support. On the other hand, in the era of globalization, where international economic relations are increasing over time and interdependencies among big countries are crucial, comparative analysis of pro-environmental preferences, decisions and behaviours among people from different countries could be useful in the field of climate policy. However, given a wide range of behavioural climate change mitigation options that have been considered in this work, it should be noted that possible policy strategies might differ depending on the corresponding type of behaviour and require a comprehensive overview.

According to Steg and Vlek (2009), possible intervention can be targeted on the relevant factors behind the corresponding behaviour. On the one hand, when behaviour is strongly associated with attitudes, changes in attitude towards particular pro-environmental behaviour should be promoted. On the other hand, when contextual factors inhibit particular behaviour, policy makers could try to remove those barriers.

Furthermore, the existing literature divides possible intervention strategies into antecedent and consequence strategies as well as into informational and structural strategies (Steg and Vlek 2009). While antecedent interventions influence one or more determinates prior to the performance of behaviour through raising problem awareness, informing about choice options and announcing the likelihood of positive and negative consequences, consequence strategies assume that the presence of positive or negative consequences will influence behaviour and are thus aimed at changing the consequences following behaviour. In order to make pro-environmental behaviour more attractive, consequence strategies use rewards and feedback, whereby feedback attaches positive consequences to the corresponding behaviour by giving information about the outcome associated with it (Abrahamse et al. 2005).

Assuming that new knowledge results in changes in attitudes, which in turn affects behaviour, similar to antecedent strategies, informational strategies aim at changing perceptions, motivations, knowledge and norms in order to heighten the awareness of environmental problems. These strategies make use of information and awareness-raising campaigns as well as persuasive communication, whereby social marketing approaches, in which the information is tailored to the needs, wants and perceived barriers of individual segments of the population are identified to be the most suitable instruments (Steg and Vlek 2009).

Considering the case when pro-environmental behaviour is rather costly due to external barriers, structural strategies might be used in order to change circumstances under which behavioural choices are made (Steg and Vlek 2009). Structural strategies use financial incentives, such as taxes, subsidies, credits and rebates, to encourage pro-environmental behaviour (Stern 1992; Steg and Vlek 2009; Abrahamse et al. 2005).

Taking into account the derived determinants of behavioural climate change mitigation measures considered in this book, all activities might benefit from the mobilization of an adequate informational strategy.

In the case of individual preferences regarding renewable energy support, elderly individuals are less likely to support renewable energy. Since the descendants of elderly people might benefit from renewable energy, policy makers should appeal to the altruistic values of elderly individuals by making use of information and awareness-raising campaigns. As for young individuals, informational strategies might help to contribute to a higher perception of positive long-term effects caused by renewable energy support.

Taking into account the empirical finding that natural disasters have no effect on the impact of environmental values on individuals' choice in favour of environmental protection within the low-income sub-sample, individuals from developing countries might not causally attribute their experience with extreme weather events to climate change (Van der Linden 2015). In this case, there is a need for basic information provision in order to overcome the lack of knowledge of the scientific consensus on climate change.

Concerning daily energy-saving behaviour, empirical results indicate that environmental motivations significantly increase the probability of preforming daily energy-saving activities among respondents in Germany and the Netherlands. In order to further stimulate individuals' motivations to perform daily energy-saving behaviours more frequently, informational strategies in form of information and awareness-raising campaigns might be employed (Steg and Vlek 2009). Considering the empirical results derived within the Belgian sub-sample, where marginal effects of environmental motivations are either not significantly different from zero or even decrease the probability of performing daily energy-saving behaviour, on the one hand, monetary rewards and financial incentives might serve as an extrinsic driving force to save energy (Abrahamse et al. 2005). On the other hand, consequence interventions might be useful, since they could make daily energy-saving activities more attractive by attaching positive consequences to it. Thus, using feedbacks households are able to obtain information about their energy consumption and can associate their energy-saving behaviour with positive outcomes (Abrahamse et al. 2005).

However, Collins et al. (2003) state that information does not necessarily lead to increased awareness, and increased awareness does not necessarily lead to action. Since information alone rarely changes individuals' behaviour and decisions, information provision must be supported by other approaches (Collins et al. 2003). According to Staddon et al. (2016) and Steg et al. (2015), intervention strategies are most effective if they are able to evoke cognitive dissonance between people's reported attitudes or values and behaviour. For this purpose, the '4 E's' model of behavioural change developed by Defra (2005) represents a comprehensive approach, which assists the UK government in developing strategies by attempting to include all possible factors which are necessary to change behaviour. The suggested strategy focuses on the need to enable, encourage, exemplify and engage people in the move toward sustainability. London's congestion charge can serve as an example of how the '4 E's' model of behavioural change functions. A combination of charging motorists together with an increased provision of buses was introduced with a huge amount of accompanying publicity. This resulted in behavioural change and contributed to a 30% reduction in congestion (Defra 2005).

The European Commission (DG Environment) also recognizes that policy interventions need to address the multiple drivers of behaviour in a coherent way, since there is a set of interrelated influences and a vast array of different contexts in which decisions are made (Umpfenbach 2014). Thus, designing policies intended to influence individuals' behaviour requires a deeper understanding of how people make these decisions and what influences them. Besides classical policy measures such as regulatory, economic, informational and behavioural tools, there is strong evidence-based support for the use of a mixture of different measures to encourage pro-environmental behaviour and preferences (Sonigo et al. 2012). For instance, it is recommended to identify attitudinal groupings since the variety of different attitudes may lead to different responses to policy from different groups (Umpfenbach 2014). Indeed, existing research indicates potential in targeting initiatives for certain segments of the consumer population, which can be defined by socio-demographic factors (OECD 2011; Söderholm 2010). Furthermore, given the strong role that habits and prompts from physical surroundings play in shaping behaviours, measures that contribute to changes in the infrastructure where behaviour takes place appear to be key for facilitating behavioural change. Taking insights from institutional economics, the decisions individuals make are strongly associated with social institutions. Thus, policy instruments targeting institutional arrangements may be also useful in facilitating pro-environmental behaviour (Umpfenbach 2014).

Although the considered climate change mitigation measures occur at the individual level, environmental issues are global, both in their essence and scale of action, and thus require national and supranational or indeed global policy options (Pereira 2015). Because local action takes place in the context of broader national frameworks, supportive supranational, national and regional policies and incentives are needed. Building environmental goals and incentives into national policies, establishing suitable policy frameworks and minimum standards, the creation of partnerships between cities and national governments as well as providing exchange platforms for global cooperation are important enabling conditions for tackling climate change also at the individual level (OECD 2014).

5.2 Limitations and Options for Future Research

The findings and implications of the studies involved in this work are subject to a number of limitations, which also need to be acknowledged.

Taking into account the cross-sectional data design of the DECC Public Attitudes Tracking survey, the World Values Survey and the survey data collected within the project "Energy Efficiency of Households in Cities: A Multi-method Analysis", the research dictates extreme caution in the interpretation of correlations as causal relationships. One option for identifying causal effects under weaker assumptions compared to cross-sectional data could be to use panel data, where the time ordering of events is known and where it is possible to investigate how a specific event changes the outcome. Another way to obtain the causal mechanism underlying the observed correlation is to make use of instrumental variables (Dienes 2016). In order to identify the causal effect of environmental motivations on individual daily energy-saving behaviour, I opt for the second option and use personal experience with one or more extreme weather event as an instrumental variable for environmental motivations. This technique requires that the instrument is related to treatment but neither directly nor indirectly related to outcome, except through the effect of the treatment itself (Rassen et al. 2009). Thus, it should be acknowledged that using as an instrument a variable that has been part of the same survey as all other variables might bias the IV estimation. A task for future research would be to identify a truly exogenous instrumental variable.

One may be confident that my contribution to a deeper understanding of behavioural CO_2 mitigation options can offer a gateway for future research. With regard to the above discussed limitations regarding the identifying causal inferences from cross sectional study design in particular, Harrison and List (2004) suggest that controlled experiments, which include laboratory and field experiments, directly construct a control group via randomization and, thus, might enable a better estimation of causal effects. Experiments provide empirical findings, which can inform policy makers, motivating the launch of new policies or changes in existing ones, and are certainly a worthwhile focus for future research.

Appendix A: Theoretical Model

The following theoretical approach is not a full theoretical model as the steady state conditions are not explicitly considered—but an augmented model could indeed include this aspect. As regards the compact model presented, I have drawn on my analysis in the EIIW discussion papers No. 202 and 245 (Udalov 2014, 2018) which should give some useful insights to the reader interested in theoretical considerations along the line of an OLG model. An Overlapping Generations Model is an adequate theoretical framework since from an environmental policy perspective the interaction of young generations and the elderly is crucial in most OECD countries.

Although there are numerous theoretical contributions which analyse environmental policy using the overlapping generations framework (Howarth and Norgaard 1992; Babu et al. 1997; John et al. 1995; John and Pecchenino 1994; Ono and Maeda 2001; Ono 2005; Mariani et al. 2009; Tubb 2011), to the best knowledge of the author, the existing literature has not paid sufficient attention to investigating the politico-economic voting outcome regarding the level of renewable energy support. This theoretical model makes use of the overlapping generations framework suggested by John and Pecchenino (1994). However, in contrast to John and Pecchenino (1994), where individuals make investments in environmental maintenance, individuals' decisions regarding their contributions to renewable energy support is investigated. For this purpose a Cobb-Douglas production function with energy as an additional input is used, whereby the energy price level of the small open economy considered is assumed to be exogenous. Since old and young individuals live in the same time period, they have to decide simultaneously on their preferred level of renewable energy support. The main goal of this theoretical model is to identify possible effects of renewable energy support on different population groups.

V. Udalov, *Behavioural Economics of Climate Change*, SpringerBriefs in Climate Studies, https://doi.org/10.1007/978-3-030-03532-7

Individuals

Following John and Pecchenino (1994), the population consists of two groups, workers and retirees. At each time period *t*, a new generation appears. Each generation lives for two periods and is composed of *L* identical individuals. Workers are born in the period *t* and are denoted as L_t. Older individuals are born in the period *t−1* and denoted as L_{t-1}. There are two generations alive in any one period, the period in which they overlap.

Young individuals are endowed with one unit of labour which they supply to firms inelastically. Each agent obtains wages. Working individuals allocate their income between current consumption (c_t), current savings (s_t) and renewable energy support (m_t). Thus, the budget constraint for a young agent in the period *t* is

$$w_t = c_t + s_t + m_t. \tag{A.1}$$

Agents face a trade-off between consumption and renewable energy support. When old, individuals consume the return from savings (s_t) and support renewable energy. The budget constraint for an old individual born in the period *t* is

$$c_{t+1} = (1 + r_{t+1})s_t - m_{t+1}. \tag{A.2}$$

Individuals born in the period *t* have preferences defined over consumption and environmental quality in old and young age. Benefits, which occur in the period *t* + *1*, have to be discounted at the discount rate *δ*. According to Ono (2005), these preferences are represented by the following utility function:

$$U_t = \ln c_t^1 + \ln Env_t + \frac{1}{(1+\delta)}\left(\ln c_{t+1}^2 + \ln Env_{t+1}\right), \tag{A.3}$$

where Env_t describes the environmental quality in the period *t* and Env_{t+1} defines the environmental quality in the period *t* + *1*.

Furthermore, individuals are assumed to be non-altruistic, which implies that the elderly do not care for the young and the young do not care for the elderly.

Firms

The firm produces a homogeneous good, using capital *(K)*, labour *(L)* and energy *(E)* in each period. The neoclassical production function is given by:

$$Y_t = K_t^{\alpha} L_t^{\beta} E_t^{1-\alpha-\beta}. \tag{A.4}$$

According to Bollino and Micheli (2011), energy *(E)* is produced, using two imperfect substitutes, namely fossil fuels *(FE)* and renewables *(RE)*:

$$E_t = FE_t^{\gamma}(\sigma m_{t-1} RE_t)^{1-\gamma}, \tag{A.5}$$

whereby renewable energy support m_{t-1} from the previous period increases the amount of renewable energy as an input factor and σ denotes the effectiveness of renewable energy support.

The profit of the firm in the period t is

$$\pi_t = p_t K_t^{\alpha} L_t^{\beta} E_t^{1-\alpha-\beta} - w_t L_t - r_t K_t - p_t^E E_t. \tag{A.6}$$

Assuming a small open economy, which faces given wages (w_t), interest rate (r_t) and energy prices (p_t^E), each firm chooses labour (L_t), capital (K_t) and energy (E_t) to maximize its profits. Thus, the first-order conditions are

$$r_t = \alpha p_t K_t^{\alpha-1} L_t^{\beta} E_t^{1-\alpha-\beta}, \tag{A.7}$$

$$w_t = \beta p_t K_t^{\alpha} L_t^{\beta-1} E_t^{1-\alpha-\beta}, \tag{A.8}$$

$$p_t^E = (1-\alpha-\beta) p_t K_t^{\alpha} L_t^{\beta} E_t^{-\alpha-\beta}, \tag{A.9}$$

Where (A.7), (A.8) and (A.9) state that the firm hires labour, capital and energy until the marginal products equal the factor prices. Due to the assumed condition of perfect competition, these conditions imply factor markets clearing.

Environmental Quality

According to John and Pecchenino (1994) and Ono and Maeda (2001), environmental quality is reduced by aggregate consumption but can be improved by renewable energy support. This mechanism is expressed as the formula:

$$Env_{t+1} = Env_t - \omega c_t + \pi m_t, \tag{A.10}$$

where Env_t is the quality of the environment in the period t. The term ωc_t, is the degradation of the environment as a result of the consumption in the period t, while πm_t measures environmental improvement as a result of renewable energy support.

Voting

The two groups of individuals vote on the level of contributions to the renewable energy support m_t by maximizing the corresponding utility function with respect to m_t. Thus, the maximization problem faced by young individuals corresponds to

$$\max U_t^{young} = \ln c_t^1 + \ln Env_t + \frac{1}{(1+\delta)}\left(\ln c_{t+1}^2 + \ln Env_{t+1}\right), \tag{A.11}$$

subject to

$$c_t = w_t - s_t - m_t,$$
$$c_{t+1} = (1 + r_{t+1})s_t - m_{t+1},$$
$$r_{t+1} = \alpha p_{t+1} K_{t+1}^{\alpha-1} L_{t+1}^{\beta} E_{t+1}^{1-\alpha-\beta},$$
$$E_{t+1} = FE_{t+1}^{\gamma}(\sigma m_t RE_{t+1})^{1-\gamma},$$
$$Env_{t+1} = Env_t - \omega c_t + \pi m_t.$$

Inserting the above constraints into (A.11), the corresponding utility function of young individuals can be derived as:

$$\begin{aligned} U_t^{young} = & \ln(w_t - s_t - m_t) + \ln Env_t \\ & + \frac{1}{(1+\delta)}\ln\left(\left(1 + \alpha p_{t+1} K_{t+1}^{\alpha-1} L_{t+1}^{\beta}\left(FE_{t+1}^{\gamma}(\sigma m_t RE_{t+1})^{1-\gamma}\right)^{1-\alpha-\beta}\right)(w_t - c_t - m_t) - m_{t+1}\right) \\ & + \frac{1}{(1+\delta)}\ln(Env_t - \omega c_t + \pi m_{tt}). \end{aligned} \tag{A.12}$$

In order to determine the optimal level of m_t^{young}, the above function has to be differentiated with respect to renewable energy support:

$$\frac{\partial U_t^{young}}{\partial m_t} = -\frac{1}{c_t^1} + \frac{1}{(1+\delta)}\left(\frac{(1-\gamma)(1-\alpha-\beta)\frac{1}{m_t} r_{t+1} s_t - (1 + r_{t+1})}{c_{t+1}^2}\right) + \frac{1}{(1+\delta)}\frac{\pi}{Env_{t+1}} = 0 \tag{A.13}$$

Considering Eq. (A.13), renewable energy support affects the utility function of young individuals through four channels. In the period *t*, there is a negative effect $-1/c_t^1$ caused by the negative impact of m_t on consumption. In period *t* + *1*, young individuals face three effects. According to (A.10), there is an environmental improvement $\pi/(1+\delta)Env_{t+1}$ in period *t+1* as a result of renewable energy support. However, renewable energy support has an ambiguous effect on consumption in period *t* + *1*:

$$\frac{1}{(1+\delta)}\left(\frac{(1-\gamma)(1-\alpha-\beta)\frac{s_t}{m_t} r_{t+1} - (1 + r_{t+1})}{c_{t+1}^2}\right) \tag{A.14}$$

On the one hand, according to (A.2), (A.5) and (A.7), renewable energy support increases an individual's consumption in the period $t + 1$. On the other hand, since there is a trade-off between renewable energy support and savings in the period t, an increase in renewable energy support has a negative effect on consumption in the period $t + 1$ due to (A.1) and (A.2). The overall effect of m_t on an individual's consumption in the period $t + 1$ is positive if the following inequality condition is fulfilled:

$$(1-\gamma)(1-\alpha-\beta) > \frac{(1+r_{t+1})m_t}{r_{t+1}s_t}. \tag{A.15}$$

Thus, the effect of m_t on an individual's consumption in the period $t + 1$ is positive if output elasticity of renewable energy is greater than the ratio of opportunity costs of renewable energy in the sense of lost consumption in the period $t + 1$ to income on savings.

Younger individuals will vote for a level of m_t that balances out negative and positive effects so that $\partial U^{young}/\partial m_t = 0$. Since long-term effects, which occur in the future, are discounted to their present value, the voting outcome of young individuals is sensitive to changes in the discount rate δ, which represents the individual's time preference. A higher δ increases preferences for the present and has a negative effect on the level of renewable energy support.

As regards the elderly, they cannot enjoy future improvements in the quality of the environment and possible benefits from the positive consumption effect in the period $t + 1$, since their maximization problem in period t is given by

$$\max U_t^{old} = \ln c_t^2 + \ln Env_t, \tag{A.16}$$

subject to

$$c_t^2 = (1+r_t)s_{t-1} - m_t.$$

Inserting the above constraint into the objective function, the utility function of older individuals is given by:

$$U_t^{old} = \ln\left((1+r_t)s_{t-1} - m_t\right) + \ln Env_t. \tag{A.17}$$

In order to estimate the retirees' optimal level of renewable energy support, the above function has to be differentiated with respect to m_t:

$$\frac{\partial U^{old}}{\partial m_t} = -\frac{1}{c_t^2} < 0. \tag{A.18}$$

Table A.1 Summary of effects and preferred level of renewable energy support

	Old individuals	Young individuals
Consumption effect (period t)	$< \mathbf{0}$	$< \mathbf{0}$
Environmental effect (period $t + 1$)	–	$> \mathbf{0}$
Consumption effect (period $t + 1$)	–	$> \mathbf{0}$ if $(1-\gamma)(1-\alpha-\beta) > \frac{(1+r_{t+1})m_t}{r_{t+1}s_t}$
Voting preferences regarding m_t	$= \mathbf{0}$	$m_t^{young} \geq m_t^{old}$

Since renewable energy support negatively affects the consumption and utility of the retirees in the period t, they will unambiguously lose from renewable energy support and vote for a zero level of m_t.

Based on the derived results, each group in society has distinct preferences regarding the level of renewable energy support, which result in an intergenerational conflict between generations alive in the period t. The corresponding effects, which influence the preferences of population groups, are summarized in the Table A.1.

Because of the divergent preferences of the two politically active population groups, the workers and the retirees, policy choices are determined through a political process. Using a majority voting mechanism, the political voting outcome depends on the assumed size of the corresponding groups. However, Gradstein and Kaganovich (2004) states that since older individuals are always the minority, the policy preferences of the older generation will have no impact on political outcomes, if age is the only determinant of policy choices. The interests of older individuals will have no impact on political outcomes and the voting outcome will correspond to the level of renewable energy support preferred by younger individuals. That is why using a majority voting mechanism in an OLG framework is problematic. Facing this problem, Gradstein and Kaganovich (2004) argue that political parties converge to platforms that maximize the aggregate welfare of the electorate. Thus, let us suppose a government which cares about the welfare of all living individuals by maximizing the aggregate utility function the period t:

$$U_t^* = (1-\mu)U_t^{old} + \mu U_t^{young}, \tag{A.19}$$

where $(1\text{-}\mu)$ is equal to $L_{t-1}/(L_t + L_{t-1})$ and represents the share of older individuals in the total population. μ is equal to $L_t/(L_t + L_{t-1})$ and denotes the share of younger individuals in the total population.

The maximization problem corresponds to

$$\max U_t^* = (1-\mu)U_t^{old} + \mu U_t^{young}, \tag{A.20}$$

subject to

$$c_t^2 = (1+r_t)s_{t-1} - m_t,$$

$$c_t^1 = w_t - s_t - m_t,$$

$$c_{t+1}^2 = (1 + r_{t+1})s_t - m_{t+1},$$

$$r_{t+1} = \alpha p_{t+1} K_{t+1}^{\alpha-1} L_{t+1}^{\beta} E_{t+1}^{1-\alpha-\beta},$$

$$E_{t+1} = FE_{t+1}^{\gamma}(\sigma m_t RE_{t+1})^{1-\gamma},$$

$$Env_{t+1} = Env_t - \omega c_t + \pi m_t.$$

Substituting the above constraints into (A.20) and building the first derivative of U_t^* with respect to m_t, the following first-order condition is obtained:

$$\frac{\partial U_t^*}{\partial m_t} = -(1-\mu)\frac{1}{c_t^2} + \mu\left[-\frac{1}{c_t^1} + \frac{1}{(1+\delta)}\left(\frac{(1-\gamma)(1-\alpha-\beta)\frac{s_t}{m_t}r_{t+1} - (1+r_{t+1})}{c_{t+1}^2}\right) + \frac{L_t}{L_t+L_{t-1}}\frac{1}{(1+\delta)}\frac{\pi}{Env_{t+1}}\right] = 0 \quad \text{(A.21)}$$

The aggregate welfare is affected by an increase in m_t through five channels. On the one hand, an increase in m_t decreases the consumption of older and younger agents in the period *t* because of the trade-off between renewable energy support and consumption. On the other hand, in the long run, an increase in m_t improves environmental quality, but also has an ambiguous effect on consumption in the period *t* + *1*. These effects are faced by young individuals who benefit from future environmental improvements and face an unclear effect of renewable energy support on long-term consumption.

In order to choose an optimal level of m_t, negative and positive effects have to be balanced out, implying that $\partial U_t^*/\partial m_t = 0$. Since government takes into account the interests of both groups, the actual voting outcome is situated between the voting preferences of younger and older individuals. The key element, which influences the actual level of renewable energy support, is the proportion of old *(1-μ)* and young individuals *(μ)*. A growth in the proportion of elderly individuals in the population increases the pressure on political representatives to choose a lower level of renewable energy support, as older individuals unambiguously lose from an increase in renewable energy support. An increase in the proportion of older individuals can be explained by population aging. An opposite effect can be seen when $\mu = L_t/(L_t+L_{t-1})$ grows and increases the political power of younger individuals, forcing the representative government to choose a higher level of renewable energy support (e.g. immigration could be such a mechanism, assuming that the average age of immigrants is below that of the host country's population). This result goes in line with Tubb (2011) who states that aging increases the political pressure on the public planner to tilt the composition of public spending in favour of a transfer payment to the elderly.

Conclusion

This theoretical model investigated the voting behaviour of different population groups regarding their renewable energy support. Based on the derived results, it is possible to identify the following effects on individuals that are caused by renewable energy support: Due to a trade-off between renewable energy support and consumption, there is a negative consumption effect in the short run. In the long run, renewable energy support improves environmental quality. However, renewable energy support has an ambiguous effect on long-term consumption. On the one hand, there is a trade-off between renewable energy support and savings, so that an increase in renewable energy support has a negative effect on future consumption. On the other hand, renewable energy support has a positive impact on the amount of produced energy, which in the long run increases consumption. While the short-term effect influences both older and younger individuals, the long-term effects influence only younger individuals. Following this line of argumentation, old individuals will unambiguously lose from renewable energy support and vote for its minimum level. In the long run, younger individuals might benefit from the positive environmental effect and an ambiguous consumption effect. Thus, based on the derived results, there is an intergenerational conflict between older and younger generations arising from different preferences regarding renewable energy support.

The limitation of the theoretical model is the assumption that there are no altruistic links between older and younger individuals. Incorporating the altruistic link between older and younger individuals would imply that children or grandchildren will inherit a better world, which also makes their parents better off—the benefit being a warm glow of satisfaction rather than direct benefits from an improved environment or increased consumption. Although allowing altruism would increase the preferred level of renewable energy support, it would not influence the presence of the derived short- and long-term effects caused by renewable energy support. However, it should be acknowledged that incorporating altruism would indeed enrich the model. In order to address the assumption that the energy price level is exogenous, the modelling of energy markets is another task for future research as well.

Appendix B: Survey Questionnaire "Energy Efficiency of Households in Cities"

Socio-Demographic

What is your gender?	Male/Female
What is your age?	. . .years old
What is the highest educational level you achieved?	Secondary school • A levels • Vocational training • University degree or higher
What is your household net income?	Answering options varied per country for comparability
What is the size of the village or city you live in (number of inhabitants)?	<5000 • 5–20,000 • 20–50,000 • 50–100,000 • 100–1,000,000 • > 1,000,000
In which province do you live?	*Netherlands*: Drenthe • Flevoland • Friesland • Gelderland • Groningen • Limburg • Noord-Brabant • Noord-Holland • Overijssel • Utrecht • Zeeland • Zuid-Holland *Germany*: Baden-Württemberg • Bayern • Berlin • Brandenburg • Bremen • Hamburg • Hessen • Mecklenburg-Vorpommern • Niedersachsen • Nordrhein-Westfalen • Rheinland-Pfalz • Saarland • Sachsen • Sachsen-Anhalt • Schleswig-Holstein • Thüringen *Belgium*: Antwerpen • Limburg • Oost-Vlaanderen • Vlaams-Brabant • West-Vlaanderen • Henegouwen • Luik • Luxemburg • Namen • Waals-Brabant
In which city do you live?	Open question
What is your current main daily activity?	Student • Regular employee • Managing employee • Employer • Self-employed • Unemployed • Stay-at-home parent • Pensioner • Other

(continued)

V. Udalov, *Behavioural Economics of Climate Change*, SpringerBriefs in Climate Studies, https://doi.org/10.1007/978-3-030-03532-7

How many adults/children live in your household? + age (including yourself)	Number of adults and number of children + age
Do you have a paid job?	Yes + hours per week profession/No
Do you have a partner?	Yes/No
[Yes] does your partner have a paid job?	Yes + hours per week profession/No
In which type of neighbourhood do you live? (more than one answer possible)	A village/rural core • A low density neighbourhood made up detached house with gardens • A neighbourhood mainly made up terraced houses • A derelict urban area • A city centre • A social neighbourhood (social housing) • A neighbourhood made up multi-story buildings (more than 5 storeys buildings)
What is the main type of buildings in your neighbourhood?	Skyscraper (more than 15 storeys) • terraced apartment buildings (2 to 15 storeys) • Terraced houses • Semi-detached houses • Detached houses • Detached apartment blocks • Farm
What is the mean number of storeys (ground floor included) in your neighbourhood?	15 and more • 7 to 14 • 4 to 6 • 2 to 3 • Just ground floor
What is (approximately) the distance from your dwelling to the nearest city centre? (If you do not know, you can give an estimate)	…km
How do you rate the availability of the following equipment in your neighbourhood? Supermarkets Proximity shops The bus/train/ tram/metro services Services related to health (doctor, etc.). Administrative services (post, administrations, etc.) Sport/leisure centres Primary schools Secondary schools High schools and universities Public parks and green spaces The amenities dedicated to walkers and bikers (pathways, etc.)	Very good to very bad (5 point Likert scale)
How do you rate the following characteristics in your neighbourhood: The level of noise The quality of air (atmospheric pollution) The congestion/traffic jam at peak hours	Very good to very bad (5 point Likert scale)

Building and Construction Details

In which type of dwelling do you live (principal dwelling)?	House • Bungalow • Flat • Maisonette/Duplex	
Do you own the house or is it rented?	Renter • Property • Other	
When was your dwelling built?	Year (if you don't know the precise year, please indicate a decade)	
In which year did you move to your current dwelling?	Year	
When you decided to move to your current dwelling, what was the importance of the following criteria?	Location close to work place • Energy label/energy consumption of the dwelling • Surface area of the dwelling • Presence of an external space (garden terrace) • Other please specify: 1 = very unimportant, 5 = very important	
	House/Bungalow	*Flat/Maisonnette-duplex*
Type	Detached • Semi-detached • Mid-terrace • End-terrace	Converted house • Above shop or office • Low storey building (<5 storeys) • High storey building (>5 storeys)
Number of dwellings (including yours) in your building	–	Number
What is the position of your flat/maisonette in the building?	–	Basement • Ground floor • Mid floor • Top floor
Number of outside-facing walls	Exposed on all sides • Between three and four • Three • Between two and three • Two • Between one and two • One	Exposed on all sides • Between three and four • Three • Between two and three • Two • Between one and two • One
Number of storeys	Number	Number (of your own dwelling)
Total living space	Amount of m2/Don't know	Amount of m2/Don't know
The surface area of my private external space (garden, terrace, etc.)	Amount of m2/Don't know	Amount of m2/Don't know
Would you take a higher mortgage to buy a more energy efficient dwelling?	Yes, I'd like to and probably will • Yes, I'd like to but probably won't • No, I don't want to • I don't know	
[Yes I'd like to and probably will] how much?	Open question	
How well is your dwelling insulated at this moment?	Very good • Good • Poor • I don't know	
(EU) what is the current energy efficiency label of your dwelling?	I don't know • A • B • C • D • E • F • G (for NL and Belgium this is answerable. The energy efficiency rating indicated the overall efficiency rating of your home. A being the highest, G the lowest. The higher the rating, the more energy efficient it is.)	

(continued)

Please indicate the current level of insulation of your dwelling: Is the roof of your building insulated? Are the walls of your building insulated? Is the basement of your building insulated?	Yes, the level of insulation is good • Yes, but the level of insulation is low • No • I don't know • N/A
What type of glazing do you have? (more than one answer possible)	Single-glazing • Old double • New double • Triple

In Home: Heating Your Home and Electricity

What is the main form of space heating/ cooling?	Boiler • Warm air • Storage heaters • Central boiler for the entire apartment building • District heating (heating shared by several buildings) • Room heaters • Ceiling heating • Air-conditioning
What is the main heating fuel?	Natural gas • Oil • Electricity • Coal • Wood • Renewable energy (e.g. biomass)
When was the boiler installed?	Year
Boiler type	Normal • Combination boiler • Condensing boiler • Condensing combination boiler • Black boiler • Don't know
Do you have a separate boiler for hot water (e.g. close-in boiler)?	Yes • No • I don't know
What are the main heating controls?	Room thermostat (temperature control) • Programmer (time clock) • TRV's (thermostatic radiator valves) Do not select TRVs if less than 50% of the total rooms are controlled by TRVs
Can you switch off the heating?	Yes, for the whole house/apartment • Yes, for each room • No • I don't know
Is the heating system programmed (it does not run continuously but only when needed)?	Yes • No, but it only runs when needed • No, it runs continuously • I don't know
What is the amount of living space that is heated during winter?	. . .%
What do you consider a comfortable/normal temperature for your living room?	• Daytime nobody home °C • Daytime someone home °C • In the evening °C • At night °C
What do you consider a comfortable/normal temperature for your sleeping room?	• In the evening °C • At night °C
Is the electricity you buy renewable?	Yes • No • I don't know
Does your home use any 'off-peak' electricity (e.g. economy or reduced-rate tariffs during the night)	Yes • No • I don't know

(continued)

[YES] do you use a timer (e.g. to turn on a pre-loaded washer or dryer during the off-peak periods?	Yes + Number • No
Do you have a smart meter (for off-peak electricity)?	Yes • No • I don't know
Is the price of electricity higher beyond a certain point?	Yes • No • I don't know
It would be easy to use less electricity in my everyday life	Strongly agree to strongly disagree (5 point Likert scale)

On the Road: Passenger Transport

How many cars are available in your household?	0 • 1 • 2 • 3
For the respondents who have at least one car	
(In case you have 2 or more cars, please answer the following questions for the main car)	
What kind of car are you driving?	New • Second-hand • More than one former owner
What kind of car (size) are you driving?	Small • Medium • Large • Small SUV • Large SUV • Transporter
(EU) what is the energy label of your car?	Categories differ according to national legislation • I don't know
Was the energy label important for choosing your car?	Very important to not important (5 point scale)
What type of fuel does your car require?	Gasoline • Diesel • LPG • Electricity
How often do you drive a car?	Every day • 4–5 times per week • 2–3 times per week • Once a week or less • I don't know
How many kilometres do you drive per year?	Number of kilometres • I don't know
How many kilometres (average) do you drive on one trip?	Number of kilometres • I don't know
What are the costs for driving per year (operation + fuel)?	. . .Euro • I don't know
Which means of transport do you use most often?	Car • train • Bus/tram • Metro • Cycling • Car-sharing • Other (several answers possible)
For which activities do you need the car?	To go to work • To go to school/ to bring the children to school • Shopping • To leisure places • Other
I travel by car because of distances to travel	Strongly agree to strongly disagree (5 point scale)
I travel by car because there are no others alternatives (public transportation, walking, cycling, etc.)	Strongly agree to strongly disagree (5 point scale)
Would you like to use the car less?	Yes/No
[Yes] would it be easy to use the car less in everyday life?	Yes/No

(continued)

If there were more amenities in my neighbourhood, I would use the car less	Strongly agree to strongly disagree (5 point scale)
If my work place was located closer to my dwelling, I would use the car less	Strongly agree to strongly disagree (5 point scale)
In case you will buy a new car, which car would you buy?	More energy efficient car (reduced GHG emissions, reduced fuel consumption) • Bigger car • Electric car • Smaller car • Same type of car as my current one
If new cars are 10% more efficient compared to your current one, would that mean that you will buy a bigger one?	Yes • No • I don't know
If the car is 10% more efficient compared to your current one, would you drive more frequently?	Yes • No • I don't know
If the fuel is 10% cheaper compared to the current one, would you drive more frequently?	Yes • No • I don't know
If the congestion charge is introduced in the CBD, would you drive less?	Yes • No • I don't know
For the respondents who have no car	
Which means of transport do you use most often?	Train • Bus/tram • Metro • Cycling • Car-sharing • Other (several answers possible)
[Train, bus/tram, metro] what are your travel costs per month for public transportation/active commuting? (if you don't know, you can give an estimate)	.. Euro
Do you have a bus or metro card?	Yes/No
What are the main reasons why you do not own a car?	It is more convenient without car • To protect the environment • It is cheaper without car • I prefer not owning a car • If I need a car I can easily rent or borrow one • I don't have a driver's licence • Other
Do you desire to have a car?	Yes/No
[Yes] what are the main reasons why you desire a car?	It would make my life easier • To save time • To save money • Others would look up to me • Other
If you had a car, for which purposes would you use it?	To go to work • To school • For shopping • To leisure places • To go on holidays • Other
If you had a car, how often would you use it?	Every day • 4–5 times per week • 2–3 times per week • Once a week or less • I don't know
If the public transport pricing (rail transit/bus/train/taxi) is 10% cheaper compared to the current one, would you choose it more?	Yes • No • I don't know
If the public transport (rail transit/bus/train/taxi) is 10% more comfortable compared to the current one, would you choose it more?	Yes • No • I don't know
If I am going to buy a car, I would choose a new-energy car	Strongly agree to strongly disagree (5 point scale) • N/A, I would not buy a car

(continued)

For ALL respondents who have a job, and whose partner has a job (questions about partner): home to work commute			
Destination	Distance from dwelling to... If distance is shorter than 1 km, please indicate 1	Frequency: How many round trip per week?	With which means of transportation do you usually travel the greatest distance to your job?
Workplace 1(you)	Km	Number	Train • Bus/tram • Metro • Cycling motor biking • Car • Car-sharing • Foot
Workplace 2 (partner)	Km	Number	Train • Bus/tram • Metro • Cycling motor biking • Car • Car-sharing • Foot
How many times per week, do you combine home-to-work travel with the following activities (you): – Go to shops – Bring children to school – Sport, leisure, visit		Number Number • N/A Number	
How many times per week, to you combine home-to-work travel with the following activities (partner): – Go to shops – Bring children to school – Sport, leisure, visit		Number Number • N/A Number	

Technical Energy Saving Measures

During the past years, what kind of actions were done to improve your house/to save energy? (H = heat; E = electricity; T = transport)		
H—House insulation (attic, wall, windows) E—Solar PV panels E—Energy efficient boiler (heating system)/air conditioner (CH)/furnace (B) E—Compact fluorescent light bulbs (CFLs) T—More efficient car T—Electric car	Yes • No • I don't know • Not applicable (because it is not something for me to decide)	[Yes] → done by: you/previous owners

(continued)

Respondents who said they had not taken an action or they didn't know received a follow-up question	
Are you considering any of the following retrofitting works over the next 12 months?	
Install new insulation in your house (attic/walls/windows) Install solar PV panels Purchase an energy-efficient boiler/air conditioner (CH)/furnace (B) Change light bulbs in your home to high CFLs Purchase a more efficient car Purchase an electric car	Yes, I'd like to and probably will • Yes, I'd like to but probably won't • No, I don't want to • I don't know
Questions on the thresholds to energy-efficiency actions were asked only if respondents who said (1) they had not undertaken the efficiency action or didn't know. Please check all of the reasons below that apply to you and select the three most important reasons	
There are many reasons why people don't {insert the action}. Please indicate the three most important reasons that apply to you	• I don't know how • It is too much effort • I am too busy • I cannot afford the investment costs • I could afford it, but don't want to spend the money • Someone else in my home would object • I don't care about my energy consumption • I don't care about the environment PV: • Energy yield is too low • Fear of gained promised efficiency • Visual representation • It takes too long to recoup the expenditure CFLs: • I don't like them • Too expensive Fuel efficient car: • I don't need a new car • They are not powerful enough • They are too small • They are not safe • It takes too long to recoup the expenditure Electric car: • I don't need a new car • They are not powerful enough • I don't like them (looks) • I am worried about the resale value • I am worried about the perceived maintenance costs • I am worried about the availability of charging stations • It takes too long to recoup the expenditure

(continued)

For each of the actions that respondents said (1) they had already done or (2) they intend to do in the coming year, they were asked their decisive factor or motivation to do	
(1) why did you {insert the action} or (2) why do you want to {insert the action}. Please indicate the most important reasons that apply to you	• It saves me money • It improves my comfort/living conditions • It improves the value of my dwelling • I had interesting financial incentives • It helps reduce global warming/avoid negative environmental impact • Someone asked me to • It's the moral thing to do • People I care about are doing it • It makes me feel good about myself • Green image • Other people approve when I do E + H: • It reduces energy consumption • It improves my house T: • It reduces fuel consumption

Behavioural Energy-Saving Measures

How often do you do the following things? (H = heat; E = electricity; T = transport)	
H—In the winter, set the thermostat to 20 °C or below H—Turn the heat down at night H—Close the windows when the heating is running E—Turn off the lights when you are not there E—Appliances on stand-by T—Walking or cycling short distances instead of driving (use car less) T—Use public transportation or car pool	Never to always (5point Likert scale)
All respondents are asked a follow-up question	
Are you considering doing the following action over the next 12 months more frequently?	
Turning down the thermostat in winter/at night Close the windows when the heating is running Turning off the lights Appliances on stand-by Walk or bike instead of drive Public transportation or car pool	• Yes, more frequently than now • No, less frequently than now • About the same as now

(continued)

Questions on the thresholds to energy conservation actions were asked only if respondents who said they continue acting in the same way over the next year or they intend to engage in the action less frequently	
There are many reasons why people don't {insert the action}. Please indicate the three most important reasons that apply to you	• I am too busy • It is hard to remember • It is not convenient • It is too much effort • I don't think it is important • Someone else in my home would object • I don't care about my energy consumption • I don't care about the environment • I am already doing this as much as I can • It would reduce my comfort Temp: • I prefer a warm house • I don't know how to set the thermostat • I cannot control the thermostat • I don't have a thermostat Light: • I am more comfortable with many lights on Walking or cycling would be a bad alternative because: • The weather makes it often not comfortable • The road is too difficult (height differences) • Travel time will be too long/the distance I travel are too far • It is not comfortable due to my health and physical condition • I prefer driving by car • Others would think it is strange if I did not use the car • I am already doing this as much as I can The public transport system is a bad alternative because: • Travel time will be too long/the distance I travel are too far • It is too expensive • There is no stop close to my home/destination • I consider public transportation is for poor people • I would not feel safe • It is not comfortable due to my health and physical condition • The weather makes it often not comfortable • I am already doing this as much as I can
Questions on the motivations to energy conservation actions were asked only if respondents who said (1) they already do often or (2) intend to do more frequently in the coming year	
Why do you want to {insert the action} more frequently. Please indicate the three most important reasons that apply to you	• It saves me money • It helps reduce global warming/negative environmental impact • It reduces energy consumption • Someone asked me to • It's the moral thing to do • People I care about are doing it • It makes me feel good about myself • Other people approve when I do Walking or cycling would be a good alternative: • It saves time • It helps to get more exercise

Willingness to Buy the New-Energy Car: Vehicle, Social and Personal Factors

Vehicle and infrastructure factors	
The vehicle type choices may be limited when buying a new-energy car	Strongly agree to strongly disagree (5 point scale)
The after-sales service may be not good after buying a new-energy car	Strongly agree to strongly disagree (5 point scale)
The investment cost is higher when buying a new-energy car	Strongly agree to strongly disagree (5 point scale)
Charging is inconvenient after buying a new-energy car	Strongly agree to strongly disagree (5 point scale)
The performance of a new-energy car may be not good	Strongly agree to strongly disagree (5 point scale)
The battery of a new-energy car may be not durable	Strongly agree to strongly disagree (5 point scale)
Social factors	
Families' and friends' suggestions are important when you buy a car	Strongly agree to strongly disagree (5 point scale)
Advertisements and salesman's suggestions are important when you buy a car	Strongly agree to strongly disagree (5 point scale)
Government propaganda and the media guide are important when you buy a car	Strongly agree to strongly disagree (5 point scale)
Personal factors	
Your basic values make you prefer to buy a new-energy car	Strongly agree to strongly disagree (5 point scale)
You prefer to buy a new-energy car because it can show your social status, personality and fashion	Strongly agree to strongly disagree (5 point scale)
Your environmental awareness make you prefer to buy a new-energy car	Strongly agree to strongly disagree (5 point scale)
Your interests to new things make you prefer to buy a new-energy car	Strongly agree to strongly disagree (5 point scale)
Your social responsibility make you prefer to buy a new-energy car	Strongly agree to strongly disagree (5 point scale)

Willingness to Reduce Energy and Beliefs About Environment Problems

Do you feel a moral obligation to reduce energy consumption?	Yes • Not really • Not at all
In which of the following decisions did environmental motivations play an important role?	
House insulation (attic, wall, windows)	I do not do such things • They play an important role • They play no role
Installing solar PV panels	I have no solar PV panels • They played an important role • They played no role
Adopting a more energy efficient boiler (heating system)/air conditioner (CH)/furnace (B)	I do not do such things • They play an important role • They play no role
Using compact fluorescent light bulbs (CFLs)	I am not using these • They played an important role • They played no role
Buying more fuel efficient car	I did not do such a thing • They played an important role • They played no role
Buying an electric car (hybrid or full battery)	I did not do such a thing • They played an important role • They played no role
Reducing energy consumption by	
Setting the thermostat to 20 °C or below in the winter	I do not do such a thing • They played an important role • They played no role
Turning the heat down at night	I do not do such a thing • They played an important role • They played no role
Closing the windows when the heating is running	I do not do such a thing • They played an important role • They played no role
Turning off the lights when you are not there	I do not do such a thing • They played an important role • They played no role
Turning appliances completely off (not leaving them on stand-by)	I do not do such a thing • They played an important role • They played no role
Walking or cycling short distances instead of driving	I do not do such a thing • They played an important role • They played no role
Using public transportation or car pooling	I do not do such a thing • They played an important role • They played no role
Recycling and waste separation	I do not do such a thing • They played an important role • They played no role
Would you say you are an environmental conscious person (if you compare yourself to others in your neighbourhood)?	Yes/No
Do you inform yourself about environmental problems and discuss such problems with others?	Yes/No
Do you feel that environmental problems are overstated?	Yes/No
Do you think your health is impaired by pollution?	Yes/No
Do you consider the internet as an important source of information on environmental progress in your country?	Yes/No

(continued)

Are you worried about: Climate change Resource scarcity Environmental degradation?	Yes/No Yes/No Yes/No
Do you think that future generations will be economically less well-off than present generations as a result of resource depletion?	Yes/No
Have you personally experienced one (or more) extreme weather event in the last few years?	Yes/No
Do you think extreme weather events are consequences of global warming and climate change?	Yes, entirely/Yes, partially/No
Do you support a recycling fee on consumer products to pay for recycling activities?	Yes/No
Do you support the ban on normal light bulbs?	Yes/No
Do you opt for energy reduction measures even if they cause extra cost?	Yes/No
After buying a new phone, would you be interested in trading in your old mobile phone if you would get a refund?	Yes/No

Consumption: See Invoice (Ask this Information in the Introduction)

The electricity consumption per year is	... kwh/year + off peak ... kwh/year/don't know
How much is your average annual electric bill?	... Euro/don't know
The heating consumption per year is	... kwh/year of m^3/year/don't know
How much is your average annual heating bill?	... Euro/don't know

Appendix 3: Survey Questionnaire "Energy Efficiency of Households in Crimea"

Appendix C: Daily Energy Saving Behaviour—Robustness Check

V. Udalov, *Behavioural Economics of Climate Change*, SpringerBriefs in Climate Studies, https://doi.org/10.1007/978-3-030-03532-7

Table C.1 Ordered probit estimation results based on the German sub-sample

	Turning heat down at night			Close windows while heating			Turning off lights when away			No appliances on stand-by		
Explanatory variables	(1)	(2)	(3)	(1)	(2)	(3)	(1)	(2)	(3)	(1)	(2)	(3)
Env_Motiv	0.156** (0.060)	0.194*** (0.059)	0.142** (0.059)	0.235*** (0.053)	0.277*** (0.051)	0.253*** (0.054)	0.110*** (0.037)	0.194*** (0.038)	0.099** (0.038)	0.161*** (0.053)	0.201*** (0.053)	0.170*** (0.053)
Female	0.100** (0.045)		0.094** (0.045)	0.146*** (0.044)		0.143*** (0.045)	0.108*** (0.034)		0.104*** (0.036)	0.075 (0.047)		0.078* (0.046)
Age	0.001 (0.002)		0.001 (0.002)	0.003** (0.001)		0.003* (0.002)	0.001 (0.001)		0.001 (0.001)	0.002 (0.002)		0.002 (0.002)
Income	−0.002 (0.016)		−0.004 (0.016)	−0.015 (0.017)		−0.017 (0.016)	−0.005 (0.011)		0.001 (0.012)	−0.011 (0.016)		−0.015 (0.015)
Education	−0.001 (0.021)		−0.012 (0.020)	−0.049*** (0.019)		−0.049** (0.019)	0.009 (0.014)		0.004 (0.014)	−0.019 (0.020)		−0.020 (0.020)
Children	0.014 (0.032)		0.009 (0.032)	−0.019 (0.028)		−0.017 (0.028)	−0.048*** (0.015)		−0.038** (0.017)	−0.031 (0.025)		−0.024 (0.024)
Family size	−0.019 (0.030)		−0.019 (0.029)	−0.013 (0.026)		−0.013 (0.026)	−0.024 (0.016)		−0.027 (0.017)	−0.070*** (0.024)		−0.066*** (0.024)
Home ownership	0.045 (0.053)		0.054 (0.052)	0.002 (0.049)		0.035 (0.050)	0.009 (0.034)		0.027 (0.034)	0.073 (0.051)		0.070 (0.049)
Heating space	−0.001 (0.001)		−0.001 (0.001)	−0.001 (0.001)		−0.001 (0.001)						
Switchable heating	0.012 (0.051)		0.033 (0.050)	0.081* (0.047)		0.084* (0.048)						
Night temperature	−0.013* (0.007)		−0.014* (0.008)	0.002 (0.007)		0.0005 (0.007)						
Day temperature				−0.024*** (0.008)		−0.020** (0.008)						
Environmental person	0.089 (0.059)		0.087 (0.059)	−0.034 (0.060)		−0.022 (0.062)	0.008 (0.039)		0.028 (0.039)	0.021 (0.062)		0.0004 (0.062)
Environmental worry	0.093 (0.067)		0.074 (0.068)	0.050 (0.062)		0.042 (0.064)	0.126*** (0.040)		0.107** (0.042)	0.081 (0.0654)		0.070 (0.066)
Province dummy	Yes	Yes	No	Yes	Yes	No	Yes	Yes	No	Yes	Yes	No
Pseudo R2	0.051	0.030	0.033	0.108	0.060	0.081	0.149	0.080	0.095	0.064	0.032	0.045
Number of observations	357	360	357	370	373	370	373	376	373	363	366	363

Average marginal effects are reported. Robust standard errors in parentheses
Significance levels: *** $p < 0.01$, ** $p < 0.05$, * $p < 0.1$

Table C.2 Ordered probit estimation results based on the Dutch sub-sample

	Turning heat down at night			Close windows while heating			Turning off lights when away			No appliances on stand-by		
Explanatory variables	(1)	(2)	(3)	(1)	(2)	(3)	(1)	(2)	(3)	(1)	(2)	(3)
Env_Motiv	0.062 (0.044)	0.110** (0.042)	0.080* (0.044)	0.257*** (0.047)	0.262*** (0.042)	0.245*** (0.049)	0.207*** (0.053)	0.197*** (0.046)	0.204*** (0.053)	0.122** (0.057)	0.139** (0.049)	0.115** (0.058)
Female	0.037 (0.037)		0.0389 (0.037)	0.018 (0.042)		0.021 (0.043)	−0.024 (0.044)		−0.021 (0.044)	0.058 (0.046)		0.050 (0.045)
Age	0.002 (0.001)		0.002 (0.001)	0.003** (0.001)		0.004** (0.002)	0.001 (0.002)		0.001 (0.002)	0.001 (0.002)		0.001 (0.002)
Income	−0.001 (0.014)		0.003 (0.014)	−0.004 (0.018)		0.007 (0.017)	−0.019 (0.016)		−0.016 (0.016)	0.001 (0.020)		−0.0004 (0.020)
Education	0.022* (0.013)		0.025* (0.013)	0.013 (0.015)		0.015 (0.015)	0.045*** (0.016)		0.044*** (0.015)	−0.006 (0.017)		−0.014 (0.016)
Children	0.001 (0.018)		0.001 (0.018)	−0.005 (0.021)		−0.011 (0.022)	0.004 (0.022)		0.002 (0.022)	0.009 (0.026)		0.008 (0.025)
Family size	0.008 (0.022)		0.001 (0.022)	−0.017 (0.024)		−0.019 (0.024)	0.005 (0.026)		0.003 (0.025)	−0.004 (0.027)		−0.006 (0.027)
Home ownership	0.031 (0.019)		0.025 (0.021)	−0.035 (0.023)		−0.032 (0.024)	−0.035 (0.022)		−0.037 (0.023)	−0.012 (0.031)		−0.012 (0.029)
Heating space	−0.001* (0.001)		−0.001** (0.001)	−0.002** (0.001)		−0.002** (0.001)						
Switchable heating	0.092*** (0.035)		0.090** (0.035)	−0.037 (0.040)		−0.040 (0.041)						
Night temperature	−0.020*** (0.007)		−0.020*** (0.007)	0.012** (0.006)		0.015** (0.007)						
Day temperature				−0.033*** (0.012)		−0.033** (0.013)						
Environmental person	0.024 (0.0407)		0.010 (0.040)	−0.018 (0.050)		−0.010 (0.053)	0.054 (0.048)		0.058 (0.049)	0.069 (0.054)		0.066 (0.056)
Environmental worry	0.049 (0.037)		0.049 (0.037)	0.022 (0.049)		0.021 (0.050)	−0.028 (0.055)		−0.036 (0.054)	−0.070 (0.052)		−0.067 (0.052)
Province dummy	Yes	Yes	No	Yes	Yes	No	Yes	Yes	No	Yes	Yes	No
Pseudo R2	0.101	0.043	0.083	0.108	0.062	0.081	0.051	0.033	0.041	0.025	0.021	0.015
Number of observations	397	426	397	402	432	402	403	434	403	352	377	352

Average marginal effects are reported. Robust standard errors in parentheses
Significance levels: *** $p < 0.01$, ** $p < 0.05$, * $p < 0.1$

Table C.3 Ordered probit estimation results based on the Belgian sub-sample

	Turning heat down at night			Close windows while heating			Turning off lights when away			No appliances on stand-by		
Explanatory variables	(1)	(2)	(3)	(1)	(2)	(3)	(1)	(2)	(3)	(1)	(2)	(3)
Env_Motiv	−0.100* (0.053)	−0.101 (0.056)	−0.169*** (0.047)	0.001 (0.052)	−0.011 (0.052)	−0.035 (0.049)	−0.014 (0.052)	0.012 (0.049)	0.009 (0.048)	0.025 (0.051)	0.075 (0.050)	−0.041 (0.046)
Female	0.026 (0.044)		0.037 (0.044)	0.084* (0.043)		0.088** (0.043)	0.077* (0.043)		0.086** (0.044)	0.085* (0.044)		0.096** (0.045)
Age	0.005*** (0.001)		0.005*** (0.001)	0.006*** (0.001)		0.005*** (0.002)	0.005*** (0.001)		0.005*** (0.001)	0.006*** (0.001)		0.006*** (0.001)
Income	−0.012 (0.015)		−0.015 (0.015)	0.009 (0.015)		0.007 (0.015)	0.022 (0.015)		0.020 (0.015)	−0.007 (0.015)		−0.007 (0.016)
Education	0.016 (0.023)		0.013 (0.024)	0.026 (0.022)		0.021 (0.022)	−0.014 (0.023)		−0.012 (0.023)	0.015 (0.024)		0.012 (0.024)
Children	−0.018 (0.027)		−0.015 (0.028)	−0.030 (0.033)		−0.027 (0.035)	−0.003 (0.027)		−0.006 (0.028)	−0.038 (0.032)		−0.032 (0.035)
Family size	−0.022 (0.024)		−0.017 (0.025)	0.038* (0.023)		0.029 (0.024)	−0.008 (0.024)		−0.006 (0.024)	−0.022 (0.028)		−0.018 (0.027)
Home ownership	0.074 (0.050)		0.064 (0.050)	0.036 (0.047)		0.037 (0.046)	−0.021 (0.049)		−0.011 (0.048)	0.053 (0.053)		0.023 (0.054)
Heating space	−0.001 (0.001)		−0.001 (0.001)	−0.001 (0.001)		−0.001 (0.001)						
Switchable heating	0.053 (0.045)		0.057 (0.046)	0.077* (0.043)		0.076* (0.043)						
Night temperature	−0.017*** (0.006)		−0.016** (0.007)	−0.002 (0.007)		−0.003 (0.007)						
Day temperature				−0.017* (0.009)		−0.014 (0.010)						
Environmental person	−0.020 (0.066)		0.009 (0.068)	−0.045 (0.062)		−0.045 (0.062)	0.065 (0.062)		0.061 (0.062)	0.041 (0.064)		0.073 (0.061)
Environmental worry	0.160** (0.068)		0.177** (0.071)	0.103 (0.063)		0.118* (0.062)	0.111 (0.068)		0.095 (0.067)	0.058 (0.067)		0.077 (0.067)
Province dummy	Yes	Yes	No	Yes	Yes	No	Yes	Yes	No	Yes	Yes	No
Pseudo R2	0.093	0.028	0.072	0.082	0.014	0.065	0.074	0.016	0.056	0.063	0.021	0.041
Number of observations	341	361	341	349	368	349	361	380	361	335	351	335

Average marginal effects are reported. Robust standard errors in parentheses
Significance levels: *** $p < 0.01$, ** $p < 0.05$, * $p < 0.1$

References

Abdullah, S. and Jeanty, P. W., 2011. "Willingness to pay for renewable energy: Evidence from a contingent valuation survey in Kenya." Renewable and Sustainable Energy Reviews 15 (6), 2974–2983.

Abrahamse, W. and Steg, L., 2009. "How do socio-demographic and psychological factors relate to households' direct and indirect energy use and savings?" Journal of Economic Psychology 30, 711–720.

Abrahamse, W., Steg, L., Vlek, C. and Rothengatter, T., 2005. "A review of intervention studies aimed at household energy conservation." Journal of Environmental Psychology 25, 273–291.

Adebo, G. M. and Ajewole, O. C., 2012. "Gender and the Urban Environment: Analysis of Willingness to pay for Waste Management Disposal in Ekiti-State, Nigeria." American International Journal of Contemporary Research 2(5).

AEE – German Agency for Renewable Energies, 2016. "Opinions on Renewables – A Look at Polls in Industrialized Countries." Renews Kompakt 29.

Akella, A. K., Saini, R. P. and Sharma, M. P., 2009. "Social, Economical and Environmental Impacts of Renewable Energy Systems." Renewable Energy 34(2), 390–396.

Angrist, J. D. and Krueger, A. B., 2001. "Instrumental Variables and the Search for Identification: From Supply and Demand to Natural Experiments." Journal of Economic Perspectives 15 (4), 69–95.

Archtnicht, M., 2012. "German car buyer's willingness to pay to reduce CO2 emissions." Climate Change 113 (3).

Babu, G. P., Kumar, K. S. K. and Murthy, N. S., 1997. "An overlapping generations model with exhaustible resources and stock pollution." Ecological Economics, 21(1): 35–43.

Bamberg, S., 2003. "How does environmental concern influence specific environmentally related behaviors? A new answer to an old question." Journal of Environmental Psychology 23, 21–32.

Bamberg, S. and Möser, G., 2007. "Twenty years after Hines, Hungerford, and Tomera: A new meta-analysis of psycho-social determinants of pro-environmental behaviour." Journal of Environmental Psychology 27, 14–25.

Batley, S., Fleming, P. and Urwin, P., 2000. "Willingness to pay for renewable energy: Implications for UK green tariff offering." Indoor Built Environment 9, 157–170.

Berglund, C. and Matti, S., 2006. "Citizen and Consumer: the Dual Role of Individuals in Environmental Policy." Environmental Politics 15 (4), 550–571.

Bergmann, A., Hanley, N. and Wright, R., 2006. "Valuing the Attributes of Renewable Energy Investments." Energy Policy 34, 1004–1014.

Bigerna, S. and Polinori, P., 2014. "Italian household's willingness to pay for green electricity." Renewable and Sustainable Energy Reviews 34,110–112.

V. Udalov, *Behavioural Economics of Climate Change*, SpringerBriefs in Climate Studies, https://doi.org/10.1007/978-3-030-03532-7

Bollino, C. A. and Micheli, S., 2011. "Sustainable growth with renewable and non-renewable energy sources." EcoMod2017, 3213.

Bostrom, A., Morgan, M. G., Fischhoff, B. and Read, D., 1994. "What do people know about global climate change? 1. Mental models." Risk Analysis 14(6), 959–970.

Brody, S. D., Zahran, S., Vedlitz, A. and Grover, H., 2008. "Examining the Relationship between Physical Vulnerability and Public Perceptions of Global Climate Change in the United States." Environment and Behavior 40 (1), 72–95.

Brounen, D., Kok, N. and Quigley, J. M., 2012. "Energy literacy, awareness, and conservation behavior of residential households." Energy Economics 38, 42–50.

Carlsson-Kanyama, A., Linden, A.-L. and Ericsson, B., 2005. "Residential energy behaviour: does generation matter?" International Journal of Consumer Studies 29, 239–252.

Cassar, A., Healy, A. and von Kessler, C., 2017. "Trust, Risk, and Time Preferences After a Natural Disaster: Experimental Evidence from Thailand." World Development 94, 90–105.

CEER, 2017. "Status Review of Renewable Support Schemes in Europe."

Ciarreta, A., Espinosa, M. P. and Pizarro-Irizar, C., 2012. "The Effect of Renewable Energy in the Spanish Electricity Market." International Conference on Future Electrical Power and Energy Systems.

Clarc, C. F., Kotchen, M. J. and Moore, M. R., 2003. "Internal and external influences on pro-environmental behaviour: Participation in a green electricity program." Journal of Environmental Psychology 23, 237–246.

Collins, C. M. and Chambers, S. M., 2005. "Psychology and situational influences on commuter-transport-mode choice." Environment and Behavior 37(6), 640–661.

Collins, J., Thomas, G., Willis, R. and Wilsdon, J., 2003. "Carrots, Sticks and Sermons: Influencing public behaviour for environmental goals." London: Demos and Green Alliances.

Costa, D. L. and Kahn, M. E., 2013. "Energy Conservation "Nudges" and Environmentalist Ideology: Evidence from a Randomized Residential Electricity Field Experiment." Journal of the European Economic Association 11, 680–702.

CRED, 2015. "The Human Cost of Weather Related Disasters: 1995–2015."

Dai, J., Kesternich, M., Löschel, A. and Ziegler, A., 2014. "Do Chinese individuals believe in global climate change and why? An econometric analysis." Joint Discussion Paper Series in Economics, 28.

Dasgupta, S., Laplante, B., Wang, H. and Wheeler, D., 2002. "Confronting the Environmental Kuznets Curve." The Journal of Economic Perspectives 16(1), 147–16.

DECC, 2015. "DECC public attitudes tracker: Wave 13." Department of Energy and Climate Change.

Defra (Department for Environment, Food and Rural Affairs), 2005. "Securing the Future: delivering UK Sustainable Development Strategy." London: Defra.

Den Butter, F. A. G. and Verbruggen, H., 1994. "Measuring the Trade-Off between Economic Growth and a Clean Environment." Environmental and Resource Economics 4(2), 187–208.

Dexter, L. A., 1964. "Communications—Pressure, Influence or Education?" In: Dexter, L. A., White, D. M. (Eds.), People, Society and Mass Communications. Collier MacMillan Ltd., London.

Diekmann, A. and Franzen, A., 1999. "The Wealth of Nations and Environmental Concern." Environment and Behavior 31, 540–549.

Dienes, C., 2016. "On the Behaviour and Attitudes of Firms and Individuals towards Resource Efficiency and Climate Change Mitigation." Josef EUL Verlag.

Dillman, D. A., Rosa, E. A. and Dillman, J. J., 1983. "Lifestyle and home energy conservation in the United States: the poor accept lifestyle cutbacks while the wealthy invest in conservation." Journal of Economic Psychology 3, 299–315.

Dorsch, M. T., 2014. "Economic Development and Determinants of Environmental Concern." Social Science Quarterly 95(4).

Drews, S., Antal, M. and van den Bergh, J. C. J. M., 2018. "Challenges in Assessing Public Opinion on Economic Growth Versus Environment: Considering European and US Data." Ecological Economics 146, 265–272.

Dunlap, R. E. and Van Liere, K. D., 1978. "The 'New Environmental Paradigm': A proposed measuring instrument and preliminary results." The Journal of Environmental Education 9(4), 10–19.

Dunlap, R. E., Van Liere, K. D., Mertig, A. G. and Jones, R. E., 2000. "Measuring endorsement of the new ecological paradigm: A revised NEP scale." Journal of Social Issues 56(3), 425–442.

Edenhofer, O., Hirth, L., Knopf, B., Pahle, M., Schlömer, S., Schmid, E. and Ueckerdt, F., 2013. "On the economics of renewable energy sources." Energy Economics 40(1), 12–23.

EEA, 2010. "Annual report 2010 and Environmental statement 2011." European Environmental Agency.

EG, 1987. "Single European Act." Official Journal of the European Communities 169/1.

Egan, P. J. and Mullin, M., 2012. 'Turning Personal Experience into Political Attitudes: The Effect of Local Weather on Americans' Perceptions about Global Warming.' Journal of Politics 74(3), 796–809.

EIA, 2016. "International Energy Outlook 2016." Independent Statistics & Analysis.

Ek, K. and Söderholm, P., 2010. "The devil is in the details: Household electricity saving behavior and the role of information." Energy Policy 38(3), 1578–1587.

EM-DAT. "The CRED/OFDA International Disaster Database." www.emdat.be, Université Catholique de Louvain, Belgium.

Eriksson, L., Garvill, J. and Nordlund, A. M., 2006. "Acceptability of travel demand management measures: the importance of problem awareness, personal norm, freedom, and fairness." Journal of Environmental Psychology 26(1), 15–26.

European Commission 2013. "The 7th Environmental Action Programme", Official Journal of the European Union 354/171.

Eurostat, 2013. "Manual for statistics on energy consumption in households." European Union, Luxembourg.

Faber, J., Schroten, A., Bles, M., Sevenster, M. and Markowska, A., 2012. "Behavioural climate change mitigation options and their appropriate inclusion in quantitative longer term policy scenarios." Delft: CE Delft.

Ferrier, B. A. and Spickett, J. T., 2007. "Natural disasters in developing countries: an environmental health perspective." Asia Pacific Journal of public Health 19, 18–24.

Fischer, C., 2008. "Feedback on household electricity consumption: a tool for saving energy?" Energy Efficiency 1, 79–104.

Franzen, A. and Meyer, R., 2010. "Environmental Attitudes in Cross-National Perspective: A Multilevel Analysis of the ISSP 1993 and 2000." European Sociological Review 26, 219–34.

Franzen, A. and Vogl, D., 2013. "Acquiescence and the Willingness to Pay for Environmental Protection: A Comparison of the ISSP, WVS, and EVS." Social Science Quarterly 94(3), 637–659.

Frederiks, E. R., Stenner, K. and Hobman, E. V., 2015. "Household energy use: Applying behavioural economics to understand decision-making and behaviour." Renewable and Sustainable Energy Reviews 41, 1385–1394.

Fürsch, M., Malischek, R. and Lindenberger, D., 2012. "Der Merit-Order-Effekt der erneuerbaren Energien - Analyse der kurzen und langen Frist", EWI.

Gallagher, K. S., 2013. "Why & How Governments Support Renewable Energy." Daedalus 142(1), 59–77.

Gardner, G. T. and Stern, P. C., 1996. "Environmental Problems and Human Behavior." Boston: Allyn and Bacon.

Gatersleben, B., Steg, L. and Vlek, C., 2002. "Measurement and determinants of environmentally significant consumer behavior." Environment and Behavior 34, 335–362.

Gradstein, M. and Kaganovich, M., 2004. "Aging population and education finance." Journal of Public Economics 88, 2469–2485.

Grasso, M., 2004. "Climate change: the global public good." Working paper del Dipartimento di Sociologia e Ricerca Sociale - Universita' degli Studi di Milano Bicocca.

Grossman, G. M. and Krueger, A. B., 1995. "Economic Growth and the environment." Quarterly Journal of Economics 110, 353–377.

Guerin, D. A., Yust, B. L. and Coopet, J. G., 2000. "Occupant Predictors of Household Energy Behavior and Consumption Change as Found in Energy Studies Since 1975." Family and Consumer Sciences Research Journal 29(1), 48–80.

Guimaraes, P., Hefner, F. and Woodward, D., 1993. "Wealth and income effects of natural disasters: An econometric analysis." Review of Regional Studies 23, 97–114.

Hage, O., Söderholm, P. and Berglund, C., 2009. "Norms and economic motivation in household recycling: Empirical evidence from Sweden." Resources, Conservation and Recycling 53, 155–165.

Hallegatte, S. and Dumas, P., 2009. "Can natural disasters have positive consequences? Investigating the role of embodies technical change." Ecological Economics 68(3), 777–786.

Hansla, A., 2011. "Value orientation and framing as determinants of stated willingness to pay for eco-labeled electricity." Energy Efficiency 4, 185–192.

Hansla, A., Gamble, A., Juliusson, A. and Gärling, T., 2008. "Psychological Determinants of Attitude towards and Willingness to Pay for Green Electricity." Energy Policy 36: 768–74.

Harrison, G. W. and List, J., 2004. "Field Experiments." Journal of Economic Literature 42(4), 1009–1055.

Henriques, S. T. and Borowiecki, K. J., 2014. "The Drivers of Long-run CO2 Emissions: A Global Perspective since 1800," Working Papers 0062, European Historical Economics Society (EHES).

Henryson, J., Hakansson, T. and Pyrko, J., 2000. "Energy efficiency in buildings through information – Swedish perspective." Energy Policy 28, 169–180.

Hersch, J. and Viscusi, K., 2006. "The Generational Divide in Support for Environmental Policies: European Evidence." NBER Working Papers 11859.

Howarth, R. B. and Norgaard, R., 1992. "Environmental valuation under sustainable development." American Economic Review 82, 473–477.

Huddart-Kennedy, E., Beckley, T. M., McFarlane, B. L. and Nadeau, S., 2009. "Rural-Urban Differences in Environmental Concern in Canada." Rural Sociology 74: 309–329.

IEA, 2016. "CO2 Emissions from Fuel Combustion."

IMF, 2014. "Proposed new grouping in WEO country classifications: Low-income developing countries." IMF Policy Paper.

Inglehart, R., 1971. "The silent revolution in Europe: Intergenerational change in post-industrial societies." American Political Science Review 65, 991–1017.

Inglehart, R., 1995. "Public Support for Environmental Protection: Objective Problems and Subjective Values in 43 Societies." Political Science and Politics 28, 57–72.

Inglehart, R., 1997. "Modernization and postmodernization: Cultural, economic and political change in 43 societies." Princeton, NJ: University press.

IPCC, 2011. "Summary for Policymakers." In: IPCC Special Report on Renewable Energy Sources and Climate Change Mitigation [O. Edenhofer, R. Pichs-Madruga, Y. Sokona, K. Seyboth, P. Matschoss, S. Kadner, T. Zwickel, P. Eickemeier, G. Hansen, S. Schlömer, C. von Stechow (eds)], Cambridge University Press, Cambridge, United Kingdom and New York, NY, USA.

IPCC, 2013. "Climate Change 2013: The Physical Science Basis. Contribution of Working Group I to the Fifth Assessment Report of the Intergovernmental Panel on Climate Change." Cambridge University Press, Cambridge, United Kingdom and New York, NY, USA.

IPCC, 2014. "Climate Change 2014: Impacts, Adaptation, and Vulnerability. Part A: Global and Sectoral Aspects. Contribution of Working Group II to the Fifth Assessment Report of the Intergovernmental Panel on Climate Change." Cambridge University Press, Cambridge, United Kingdom and New York, NY, USA.

IRENA, 2015. "Renewable Power Generation Costs in 2014."

Israel, D. and Levinson, A., 2004. "Willingness to Pay for Environmental Quality: Testable Empirical Implications of the Growth and Environmental Literature." The Berkely Electronic Press.

Ito, N., Takeuchi, K., Tsuge, T. and Kishimoto, A., 2010. "Applying threshold models to donations to a renewable electricity fund." Energy Policy 38, 1819–1825.

Jäger, J., 2007. "Was verträgt unsere Erde noch?: Wege in die Nachhaltigkeit."Forum für Verantwortung, Band 17270.

Jäger, P. and Schmidt, T., 2015. "The Political Economy of Public Investment when Population is Aging – A Panel Co-integration Analysis." Ruhr Economic Papers 557.

Jansson, J., Marell, A. and Nordlund, A., 2010. "Green consumer behaviour: determinants of curtailment and eco-innovation adoption." Journal of Consumer Marketing 27(4), 358–370.

John, A. and Pecchenino, R., 1994. "An Overlapping Generations Model of Growth and the Environment." Economic Journal 104, 1393–1410.

John, A. and Pecchenino, R., Schimmelpfennig D. and Schreft S., 1995. "Short-Lived Agents and the Long-Lived Environment." Journal of Public Economics 58, 127–141.

Jones, R. E. and Dunlap, R. E., 1992. "The social bases of environmental concern: have they changed overtime?" Rural Sociology 57(1): 28–47.

Jones, R. E. and Dunlap, R. E., 2002. "Environmental Concern: Conceptual and Measurement Issues." in R. E. Dunlap and W. Michelson (eds). Handbook of Environmental Sociology, Westport: Westwood Press.

Kaiser, F. G., Wolfing, S. and Fuhrer, U., 1999. "Environmental attitude and ecological behaviour." Journal of Environmental Psychology 19, 1–19.

Kaya, Y., 1990. "Impact of Carbon Dioxide Emission Control on GNP Growth: Interpretation of Proposed Scenarios." Paper presented to the IPCC Energy and Industry Subgroup, Response Strategies Working Group, Paris. Mimeo.

Klomp, J. and Valckx, K., 2014. "Natural disasters and economic growth: A meta-analysis." Global Environmental Change 26(1), 183–195.

Kozluk, T. and Zipperer, V., 2014. "Environmental policies and productivity growth – a critical review of empirical findings. "OECD Journal: Economic Studies 2014.

Lazzaroni, S. and van Bergeijk, P., 2014. "Natural disasters' impact, factors of resilience and development: A meta-analysis of the macroeconomic literature." Ecological Economics 107, 333–346.

Lee, E., Park, N.-K. and Han, J. H., 2013. "Gender difference in environmental attitude and behaviors in adoption of energy-efficient lighting at home." Journal of Sustainable Development 6(9), 36–50.

Leiserowitz, A., Maibach, E., Roser-Renouf, C., Feinberg, G. and Howe, P., 2012. "Extreme weather and global climate change in the American mind", Yale University and George Mason University, New Haven, CT, Yale Project on Global Climate Change Communication.

Leiserowitz, A., Maibach, E., Roser-Renouf, C., Feinberg, G., Rosenthal, S., Marlon, J. and Howe, P., 2014. "Extreme weather and global climate change in the American mind.", Yale University and George Mason University, New Haven, CT, Yale Project on Global Climate Change Communication.

Liu, W. et al., 2013. "Rural public acceptance of renewable energy deployment: The case of Shandong in China." Applied Energy 102.

Lorenzoni, I., Nicholson-Cole, S. and Whitmarsh, L., 2007. "Barriers perceived to engaging with climate change among the UK public and their policy implications." Global Environmental Change 17, 445–459.

Mariani, F., Perez-Barahona, A. and Raffin, N., 2009. "Life Expectancy and the Environment." IZA Discussion Paper 4564.

McDermott, T., 2012. "The Effects of Natural Disasters on Human Capital Accumulation." IIIS Discussion Paper 391.

Menanteau, P., Finon, D. and Lamy, M., 2003. "Prices versus quantities: choosing policies for promoting the development of renewable energy." Energy Policy, 31(8), 799–812.

Ménioux, J. and Zumsteeg, S., 2012. "Individual perceptions of climate risks: SURVEY AXA/IPSOS."
Meyer, R. and Liebe, U., 2010. "Are the affluent prepared to pay for the planet? Explaining willingness to pay for public and quasi-private environmental goods in Switzerland." Population and Environment 32, 42–65.
Moula, M. M. E., Maula, J., Hamdy, M., Fang, T., Jung, N. and Lahdelma, R., 2013. "Researching social acceptability of renewable energy technologies in Finland." International Journal of Sustainable Built Environment 2(1), 89–98.
Mozumder, P., Vasquez, W. F. and Marathe, A., 2011. "Consumers' preference for renewable energy in the southwest USA." Energy Economics 33, 1119–1126.
Murphy, R., 1994. "Rationality and Nature: A Sociological Inquiry into a Changing Relationship." Boulder: Westview Press.
Nejat, P., Jomehzadeh, F., Taheri, M. M., Gohari, M. and Majid, M. Z. A., 2015. "A global review of energy consumption, CO 2 emissions and policy in the residential sector (with an overview of the top ten CO 2 emitting countries)." Renewable and Sustainable Energy Reviews 43, 843–862.
Neumayer, E. and Plümper, T., 2008. "The Gendered Nature of Natural Disasters: The Impact of Catastrophic Events on the Gender Gap in Life Expectancy, 1981–2002." Annals of the Association of American Geographers 97(3), 551–566.
Nordlund, A. M. and Garvill, J., 2003. "Effects of values, problem awareness, and personal norm on willingness to reduce personal car use." Journal of Environmental Psychology 23, 339–347.
Norton, E., Wang, H. and Ai, C., 2004. "Computing Interaction Effects and Standard Errors in Logit and Probit Models." STATA Journal 4, 103–116.
Noy, I., 2009. "The Macroeconomic Consequences of Disasters." Journal of Development Economics 88(2), 221–231.
OECD, 2011. "Greening Household Behaviour: The Role of Public Policy." OECD Publishing, Paris:
OECD, 2014. "Cities and Climate Change." OECD Publishing, Paris.
Ono, T., 2005. "The Political Economy of Environmental Taxes with an Aging Population." Environmental & Resource Economics 30, 165–194.
Ono, T. and Maeda, Y., 2001. "Is Aging Harmful to the Environment?" Environmental and Resource Economics 20, 113–127.
Panwar, N., Kaushik, S. and Kothari, S., 2011. "Role of renewable energy sources in environmental protection: A review." Renewable and Sustainable Energy Reviews 15, 1513–1524.
Pereira, J. C., 2015. "Environmental issues and international relations, a new global (dis)order – the role of International Relations in promoting a concerted international system." Revista Brasileira de Política Internacional, 58(1), 191–209.
Peterson, T. C., Scott, P. A. and Herring, S., 2012. "Explaining Extreme Events of 2011 from a Climate Perspective." American Meteorological Society.
Poortinga, W., Steg, L., Vlek, C. and Wiersma, G., 2003. "Household preferences for energy-saving measures. A Conjoint Analysis." Journal of Economic Psychology, 24, 49–64.
Raddatz, C., 2007. "Are External Shocks Responsible for the Instability of Output in Low-Income Countries?" Journal of Development Economics 84, 155–187.
Rassen, J. A., Schneeweiss, S., Glynn, R. J., Mittleman, M. A. and Brookhart, M. A., 2009. "Instrumental variable analysis for estimation of treatment effects with dichotomous outcomes." American Journal of Epidemiology 169(3), 273–84.
Ribes, A., Zwiers, F. W., Azaïs, J. M. and Naveau, P., 2017. "A new statistical approach to climate change detection and attribution." Climate Dynamics 48(1–2), 367–386.
Rokeach, M., 1973. "The nature of human values." New York: Free Press.
Roodman, D., 2011. "Fitting fully observed recursive mixed-process models with cmp." Stata Journal, StataCorp LP 11(2), 159–206.

Running, K., 2012. "Examining Environmental Concern in Developed, Transitioning and Developing Countries: A Cross-Country Test of the Objective Problems and the Subjective Values Explanations." World Values Research 5(1), 2012.

Sabatini, F., 2012. "Who Trusts Berlusconi? An Econometric Analysis of the Role of Television in the Political Arena." Kyklos 65, 111–131.

Sardianou, E., 2007. "Estimating energy conservation patterns of Greek households." Energy Policy 35, 3778–3791.

Ščasný, M. and Urban, J., 2009. "Residential energy efficiency: a cross country empirical analysis." Paper prepared for the OECD Conference on Household Behaviour and Environmental Policy. OECD Environment Directorate. Paris.

Schneider, F., Kallis, G., and Martinez-Alier, J., 2010. "Crisis or opportunity? Economic degrowth for social equity and ecological sustainability. Introduction to this special issue." Journal of Cleaner Production 18(6), 511–518.

Schwartz, S. H., 1977. "Normative influences on altruism." In L. Berkowitz (Ed.), Advances in experimental social psychology 10, 221–279.

Schwartz, S. H., 2012. "An Overview of the Schwartz Theory of Basic Values", Online Readings in Psychology and Culture 2(1).

Schwartz, S. H. and Howard, J. A., 1981. "A normative decision-making model of altruism." In J. P. Ruston and R. M. Sorrentino (Eds.), Altruism and helping behavior: Social, personality, and developmental perspectives, Hillsdale, NJ: Lawrence Erlbaum, 189–211.

Smil, V., 2003. "Energy at the Crossroads Global Perspectives and Uncertainties." The MIT Press, Cambridge, MA, xiv + 427 pp. Book.

Söderholm, P., 2010. "Environmental Policy and Household Behaviour: Sustainability and Everyday Life." Washington DC: Earthscan.

Sonigo, P. et al., 2012. "Policies to encourage sustainable consumption." Final report prepared by BIO Intelligence Service for European Commission (DG ENV).

Staddon, S. C., Cycil, C., Goulden, M., Leygue, C. and Spence, A., 2016. "Intervening to change behaviour and save energy in the workplace: A systematic review of available evidence." Energy Research & Social Science 17, 30–51.

Steg, L. and Vlek, C., 2009. "Encouraging pro-environmental behaviour: An integrative review and research agenda." Journal of Environmental Psychology 29, 309–317.

Steg, L., Perlaviciute G. and van der Werff, E., 2015. "Understanding the human dimensions of a sustainable energy transition." Frontiers in Psychology 6 (805).

Stern, P. C., 1992. "Psychological Dimensions of Global Environmental Change." Annual Review of Psychology 43, 269–302.

Stern, P. C., 2000. "Toward a coherent theory of environmentally significant behavior." Journal of Social Issues 56 (3), 407–424.

Stern, N., 2006. "The Stern Review on the Economic Effects of Climate Change." Population and Development Review 32, 793–798.

Stern, P. C. and Dietz, T., 1994. "The value basis of environmental concern.' Journal of Social Issues 50(3), 65–84.

Stern, P. and Gardner, G., 1981. "Psychological research and energy policy." American Psychologist 36 (4), 329–342.

Stern, P. C., Dietz, T. and Guagnano, G. A., 1995. "The new ecological paradigm in social psychological context." Environment and Behavior 27, 723–743.

Stern, P. C., Dietz, T., Abel, T., Guagnano, G. A. and Kalof, L., 1999. "A value-belief-norm theory of support for social movements: The case of environmentalism." Research in Human Ecology 6 (2), 81–97.

Strömberg, D., 2007. "Natural Disasters, Economic Development, and Humanitarian Aid." Journal of Economic Perspectives 21(3), 199–222.

Sundt, S. and Rehdanz, K., 2014. "Consumer's Willigness to Pay for Green Electricity: A Meta-Analysis of the Literature." Kiel Working Paper 1931.

Swim, J. K., Clayton, S. and Howard, G. S., 2011. "Human behavioral contributions to climate change: Psychological and contextual drivers." American Psychologist 66.

Trump, D., 2017. "Statement by President Trump on the Paris Climate Accord." The White House, 01 June 2017, https://www.whitehouse.gov/the-press-office/2017/06/01/statement-president-trump-paris-climate-accord.

Tsushima, S., Tanabe, S. and Utsumi, K., 2014. "Worker's awareness and indoor environmental quality in electricity-saving offices." Building and Environment, 1–10.

Tubb, A., 2011. "Does Population Ageing Affect Government Environmental Expenditure?" Committee for Economic Development of Australia.

Udalov, V., 2014 "Political-Economic Aspects of Renewable Energy: Voting on the Level of Renewable Energy Support." EIIW Discussion Papers, 202.

Udalov, V., 2018 "Analysis of Individual Renewable Energy Support: An Enhanced Model." EIIW Discussion Papers, 245.

Udalov, V., Perret, J. and Vasseur V., 2017. "Environmental motivations behind individuals' energy efficiency investments and daily energy-saving behaviour: evidence from Germany, the Netherlands and Belgium." International Economics and Economic Policy 14(3): 481–499.

Umpfenbach, K., 2014. "Influences on consumer behaviour. Policy implications beyond nudging." Berlin: Ecologic Institute.

UNISDR, 2015. "The Human Cost of Weather Related Disasters."

Urban, J. and Ščasný, M., 2012. "Exploring domestic energy-saving: The role of environmental concern and background variables." Energy Policy 47, 69–80.

Van der Linden, S., 2015. "The social-psychological determinants of climate change risk perceptions: Towards a comprehensive model." Journal of Environmental Psychology 41, 112–124.

Van der Linden, S., Leiserowitz, A., Feinberg, G. and Maibach, E., 2014. "How to communicate the scientific consensus on climate change: Plan facts, pie charts or metaphors?" Climatic Change 126(1–2), 255–262.

Weidner, H., 1995. "25 years of modern environmental policy in Germany: Treading a well-worn path to the top of the international field", WZB Discussion Paper, FS II, 95–301.

Whitmarsh, L., 2008. "Are flood victims more concerned about global climate change than other people? The role of direct experience in risk perception and behavioral response." Journal of Risk Research 11 (3), 351–374.

Whitmarsh, L., 2009. "Behavioural responses to climate change: Asymmetry of intentions and impacts." Journal of Environmental Psychology 29, 13–23.

Whitmarsh, L., Seyfang, G. and O'Neill, S., 2011. "Public engagement with carbon and climate change: To what extent is the public 'carbon capable'?" Global Environmental Change 21 (1), 56–65.

Windahl, S. and Signitzer, B., 1992. "Using Communication Theory. An Introduction to Planned Communication." Sage Publications, London.

World Values Survey (WVS), Wave 6, 2010–2014. World Values Survey Association (www.worldvaluessurvey.org). Aggregate File Producer: Asep/JDS, Madrid, Spain.

Zelezny, L. C., Chua, P. and Aldrich, C., 2000. "Elaborating on gender Differences in Environmentalism." Journal of Social Issues 56 (3), 443–457.

Zografakis, N., Sifaki, E., Pagalou, M., Nikitaki, G., Psarakis, V. and Tsagarakis, K. P., 2010. "Assessment of public acceptance and willingness to pay for renewable energy sources in Crete." Renewable Sustainable Energy Review 14(3), 1088–1095.

MIX
Papier aus verantwortungsvollen Quellen
Paper from responsible sources
FSC® C105338

If you have any concerns about our products,
you can contact us on
ProductSafety@springernature.com

In case Publisher is established outside the EU,
the EU authorized representative is:
Springer Nature Customer Service Center GmbH
Europaplatz 3, 69115 Heidelberg, Germany

Printed by Libri Plureos GmbH
in Hamburg, Germany